A FARMER'S PRIMER ON GROWING COWPEA ON RICELAND

R.K. Pandey

International Rice Research Institute
and
International Institute of Tropical Agriculture

1987
International Rice Research Institute
Los Baños, Laguna, Philippines
P.O. Box 933, Manila, Philippines

The International Rice Research Institute (IRRI) was established in 1960 by the Ford and Rockefeller Foundations with the help and approval of the Government of the Philippines. Today IRRI is one of the 13 nonprofit international research and training centers supported by the Consultative Group on International Agricultural Research (CGIAR). The CGIAR is sponsored by the Food and Agriculture Organization (FAO) of the United Nations, the International Bank for Reconstruction and Development (World Bank), and the United Nations Development Programme (UNDP). The CGIAR consists of 50 donor countries, international and regional organizations, and private foundations.

IRRI receives support, through the CGIAR, from a number of donors including: the Asian Development Bank, the European Economic Community, the Ford Foundation, the International Development Research Centre, the International Fund for Agricultural Development, the OPEC Special Fund, the Rockefeller Foundation, the United Nations Development Programme, the World Bank, and the international aid agencies of the following governments: Australia, Belgium, Canada, China, Denmark, France, Federal Republic of Germany, India, Italy, Japan, Mexico, Netherlands, New Zealand, Norway, Philippines, Saudi Arabia, Spain, Sweden, Switzerland, United Kingdom, and United States.

The responsibility for this publication rests with the International Rice Research Institute.

ISBN 971-104-169-3

Contents

Foreword

Rice and rice-based cropping systems occupy a position of overwhelming importance in global food production. Legume crops such as cowpea fit well into these systems, helping to increase productivity by yielding more food from the same land area.

Cowpea grown either before or after rice enriches the soil, helps to break the pest and disease cycle that occurs in continuous rice cropping, and adds to farm income. Nutritionally, cowpea complements rice, adding protein to largely starchy subsistence diets. Grown for centuries in the tropics, cowpea is well adapted to prevailing environmental stresses. The crop tolerates drought and can grow on poor, even acid soils. Improved short or medium duration varieties from the International Institute of Tropical Agriculture (IITA) can be profitably fitted into a wide range of cropping systems as a food, fodder, or green manure crop requiring minimum inputs.

A Farmer's Primer on Growing Cowpea on Rice Land explains the "hows" and "whys" of cowpea culture to farmers, extension workers, students, and technicians. The Primer is patterned after *A Farmer's Primer on Growing Rice* — which has been translated into more than 30 languages — and is similarly designed for easy translation and co-publication in developing countries. The English text has been blocked off from the line drawings. The International Rice Research Institute (IRRI) will make complimentary sets of the illustrations available to cooperators, who may translate the text, strip the translations onto the illustrations, and print a translated edition on local presses.

The cowpea Primer was made possible by a collaborative project sponsored by IRRI and IITA. A companion volume is *A Farmer's Primer on Growing Soybean on Rice Land.*

Ms. Vrinda Kumble of Editorial Consultants Services, New Delhi, India, edited both the cowpea and soybean Primers.

<table>
<tr><td>M.S. Swaminathan</td><td>Lawrence Stifel</td></tr>
<tr><td>Director General</td><td>Director General</td></tr>
<tr><td>International Rice Research
   Institute</td><td>International Institute
   of Tropical Agriculture</td></tr>
</table>

The cowpea crop

The cowpea crop

Why grow cowpea

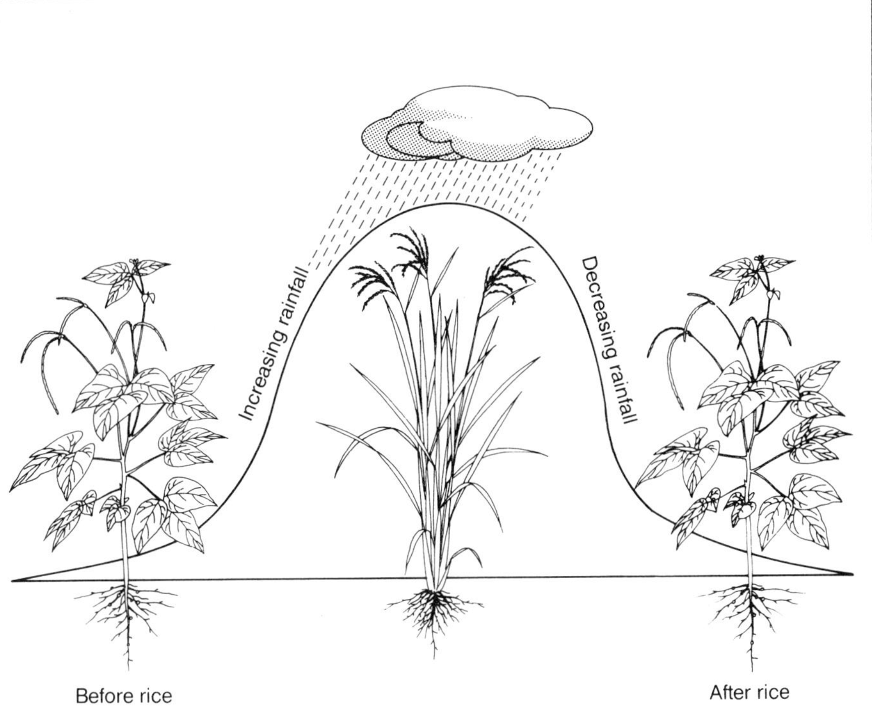

- Cowpea is a good crop to grow before or after rice.
- It can stand drought or heavy rain.
- It can grow on many kinds of soil, even acid soil, where mungbean and soybean cannot grow.

Cowpea enriches the soil

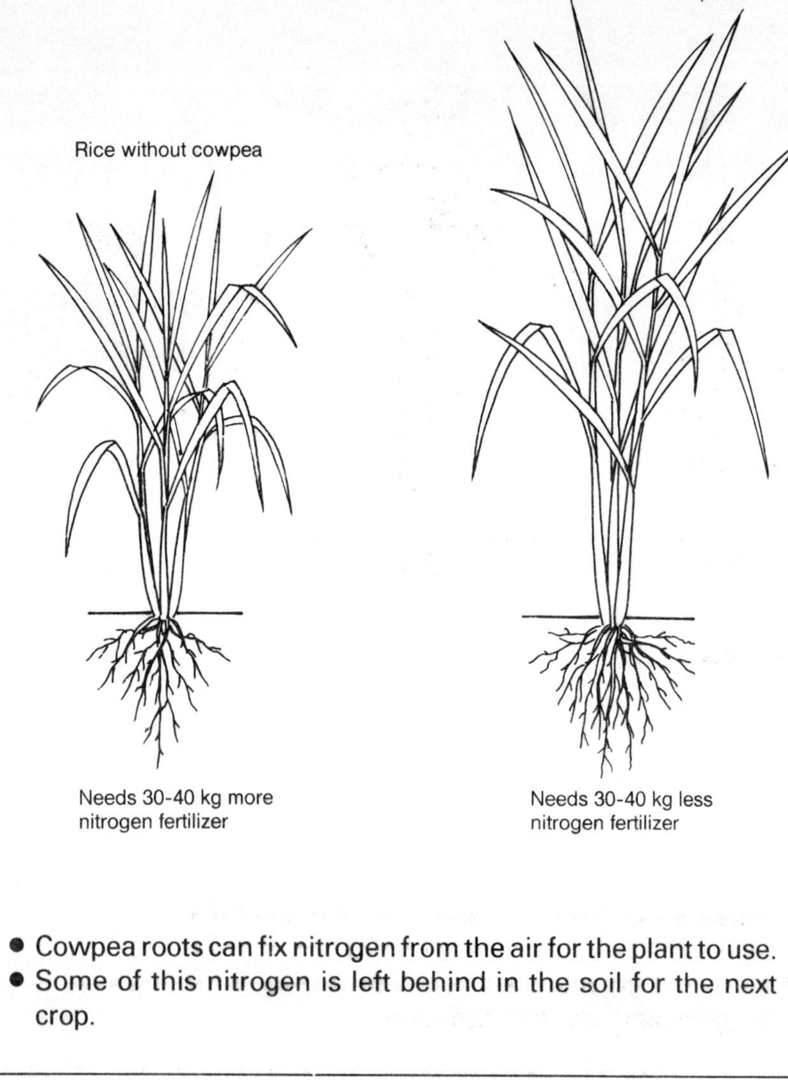

- Cowpea roots can fix nitrogen from the air for the plant to use.
- Some of this nitrogen is left behind in the soil for the next crop.

Breaks the pest and disease cycle

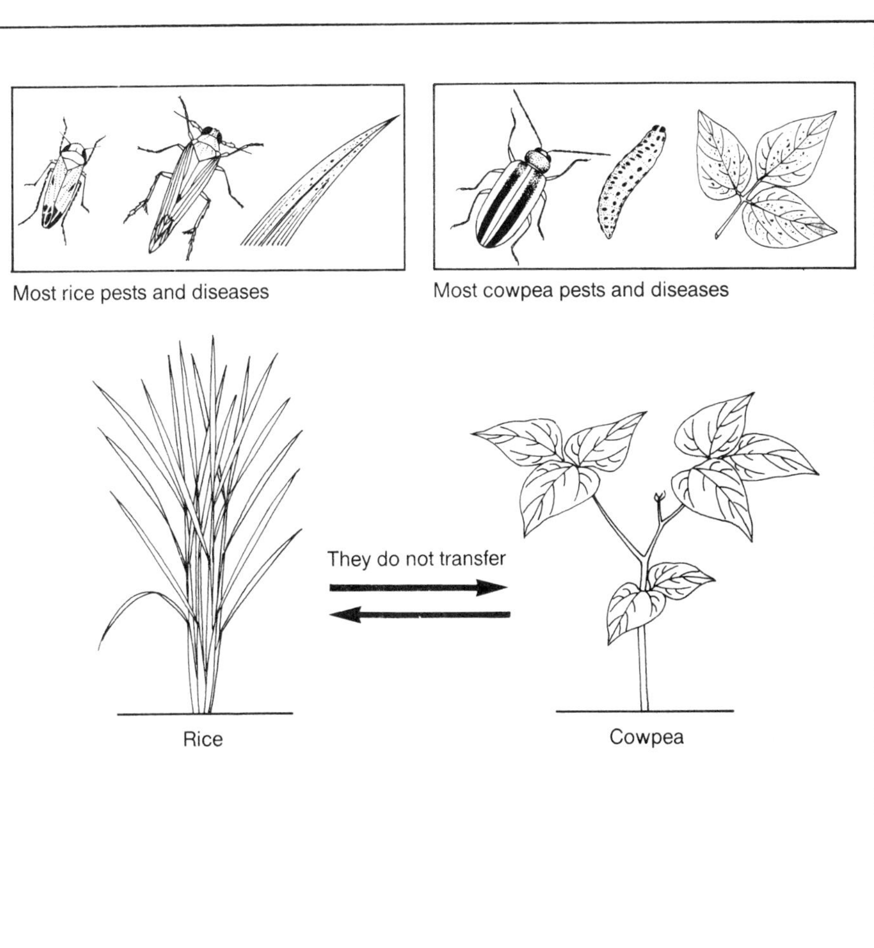

- Growing cowpea in rotation with rice breaks the pest and disease cycle for both crops because
  — most rice pests and diseases do not transfer to cowpea
  — most cowpea pests and diseases do not transfer to rice.

Adds to income

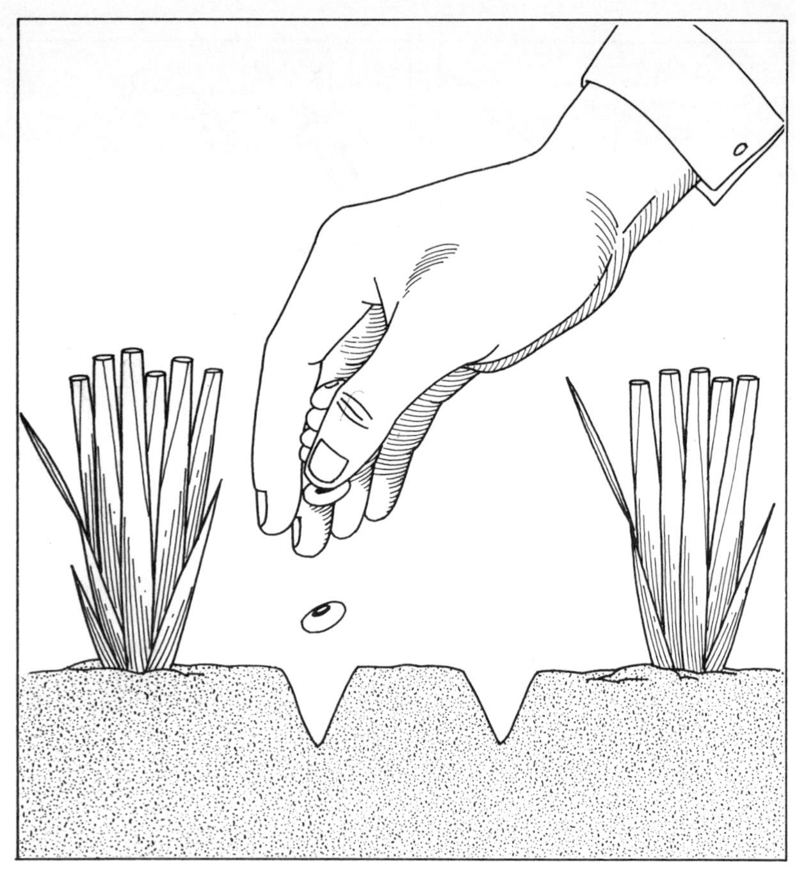

- In the off season after the rice harvest, cowpea cropping can create new jobs.

Cowpea as human food

Young leaves can be eaten as greens

Tender pods used as vegetables

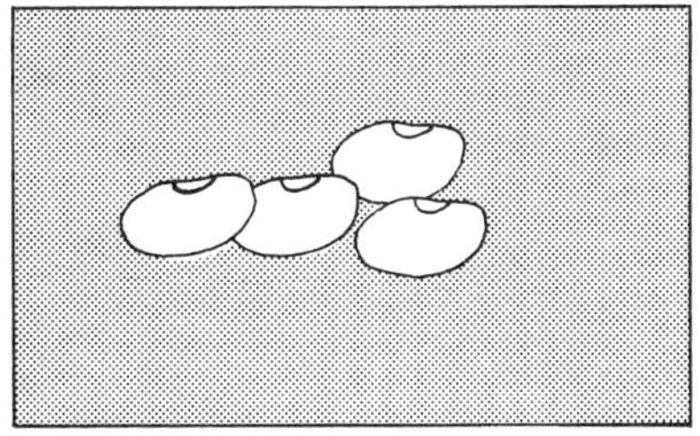

Mature seeds eaten like green peas

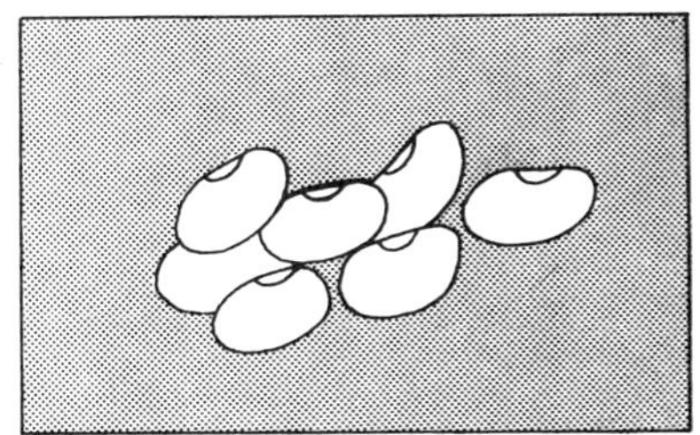

Dried seeds as a bean

- Cowpea can be eaten as greens, as a vegetable, and as a dried bean.
- Rice and cowpea eaten together make a balanced food. The nutrients lacking in each are supplied by the other.

Cowpea as fodder

- The whole plant can be used for fodder.
- Dried seed can be used for animal feed.

The cowpea plant

The cowpea plant

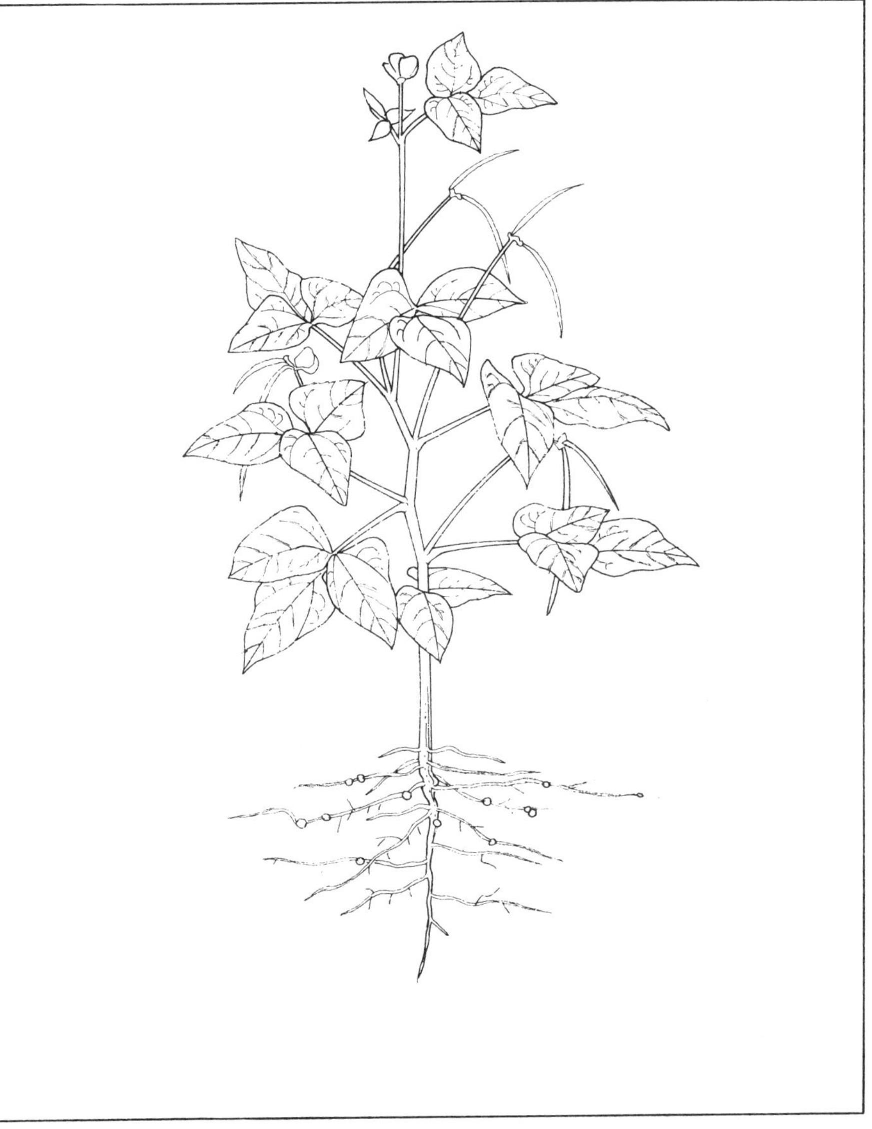

Plant types — growth habit

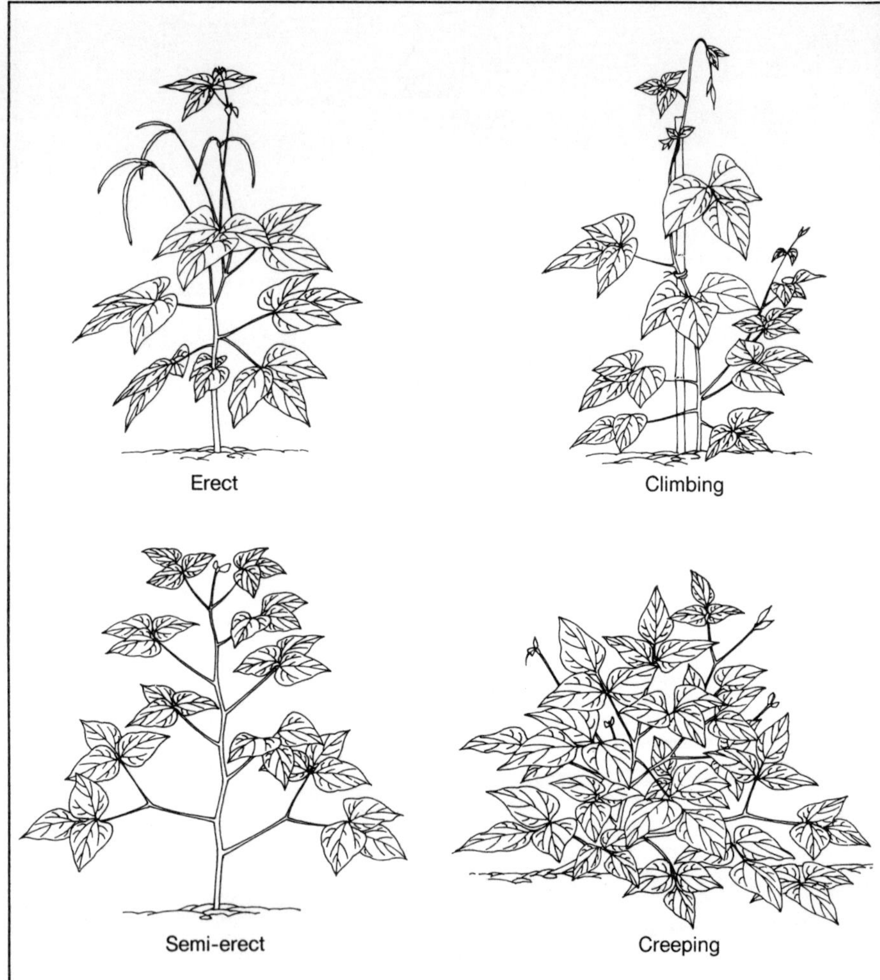

- The cowpea plant may be creeping, climbing, semi-erect, or erect.

Plant types — uneven-maturing

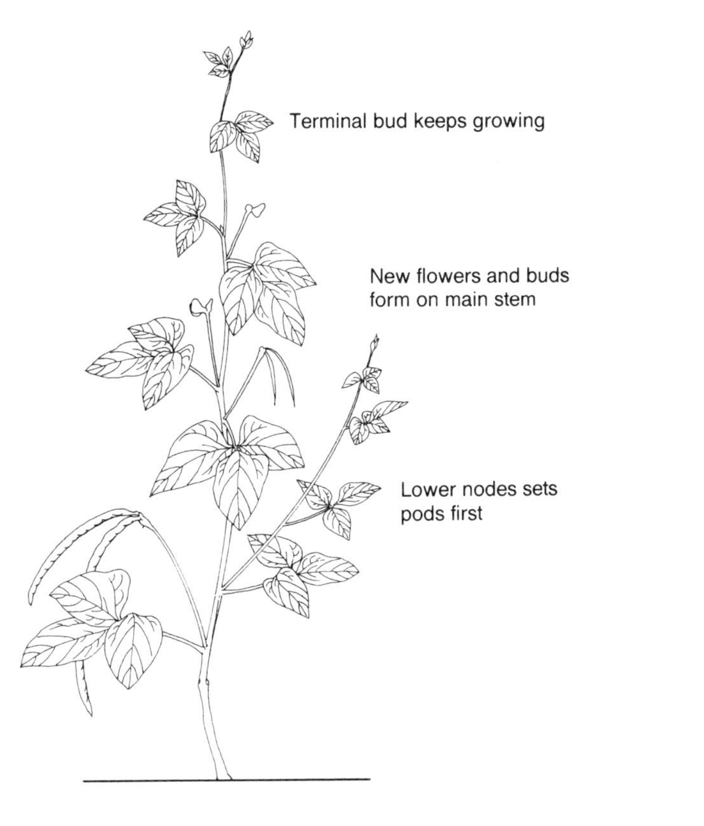

- Uneven-maturing (indeterminate) types twine or climb.
- They flower over a long period and pods do not mature at the same time.
- Rain during pod ripening may produce a new flush of flowers.

Plant types — even-maturing

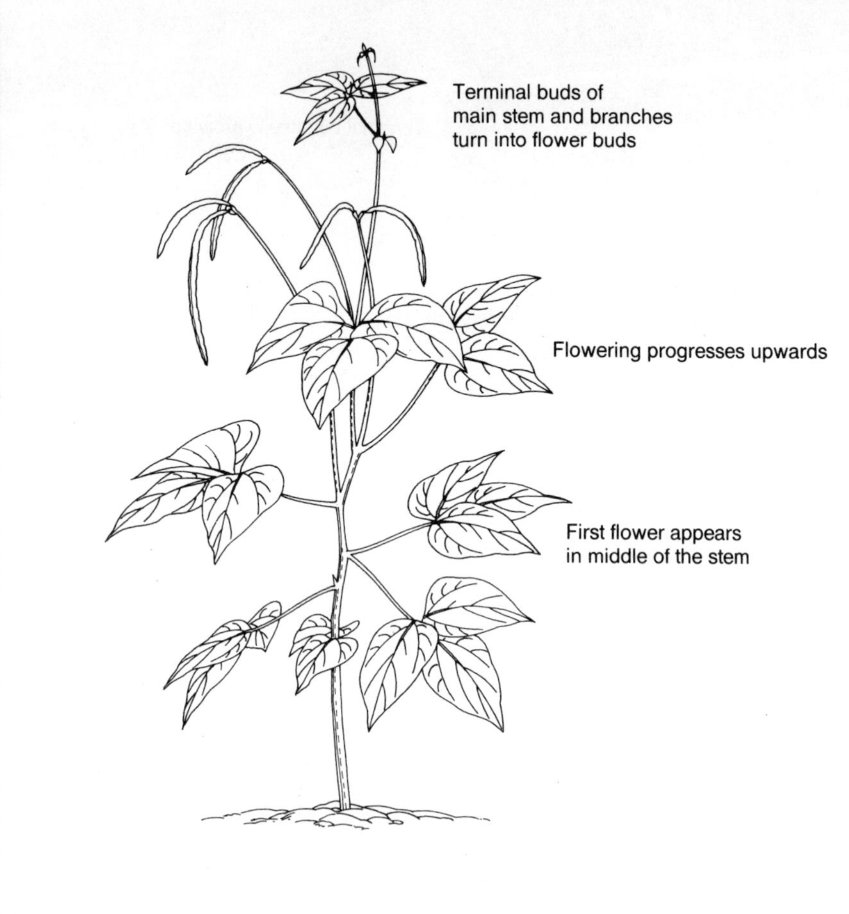

- Even-maturing (determinate) plant types grow erect.
- Most of the pods mature at the same time.

Cowpea varieties — growth duration

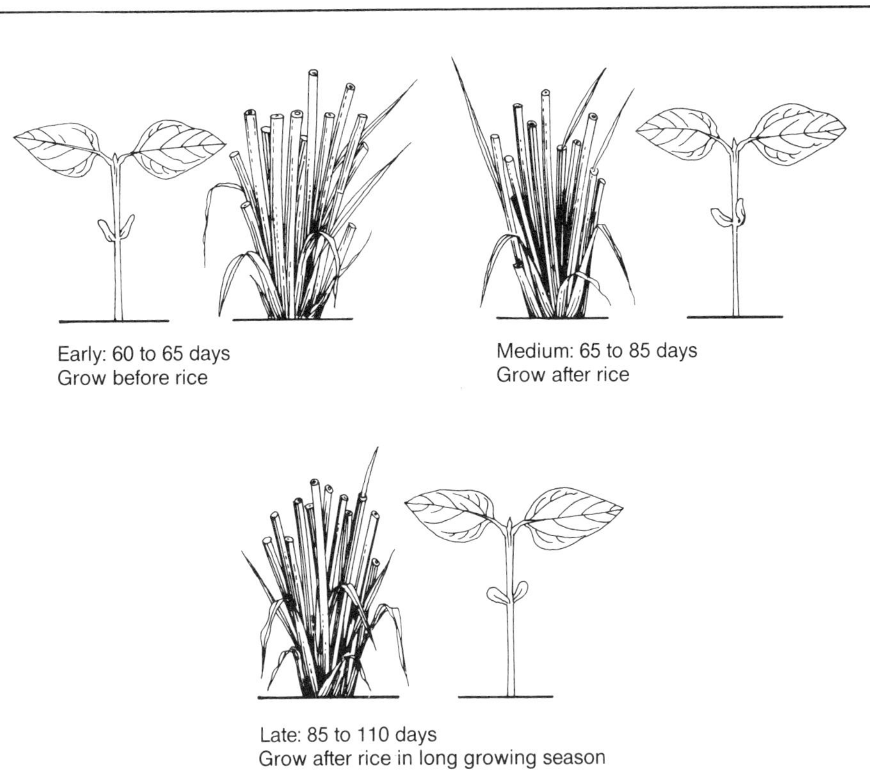

Early: 60 to 65 days
Grow before rice

Medium: 65 to 85 days
Grow after rice

Late: 85 to 110 days
Grow after rice in long growing season

- Cowpea varieties differ widely in growth habit and duration.
- Choosing the right variety to fit a cropping system gives good returns.

Cowpea varieties — uses

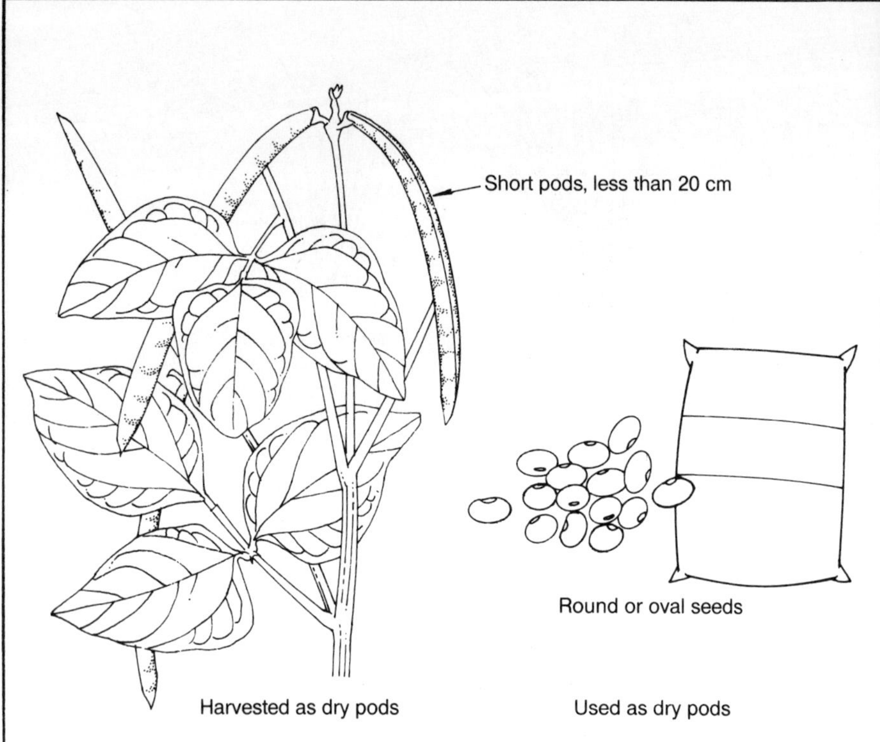

- Seed types are grown for use as dry beans.
- Cowpea seed provides nutritious human food and animal feed.

Vegetable types

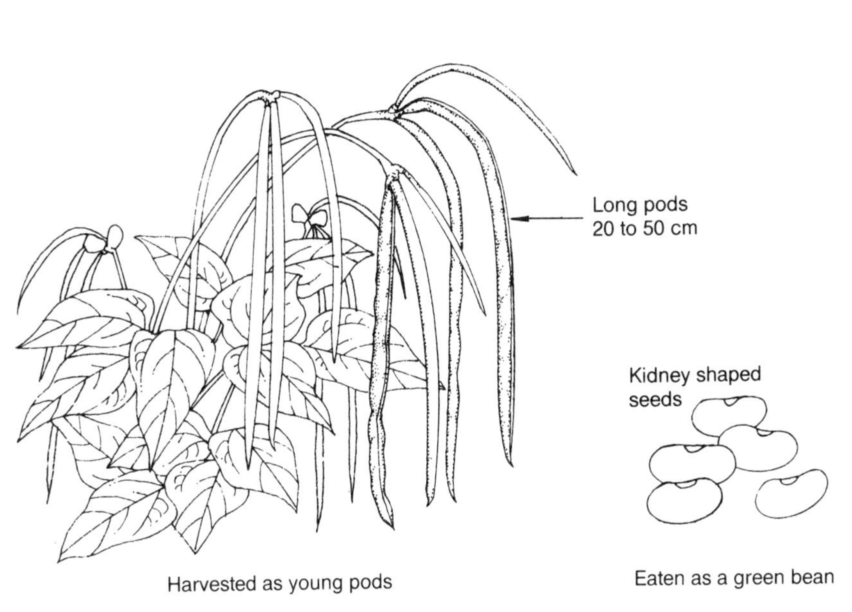

Harvested as young pods

Eaten as a green bean

- Vegetable types produce good-tasting pods that are usually longer than seed types.
- Mature seeds can also be used fresh, like green peas.

Fodder types

Whole plant harvested at flowering
or after harvest of green pods or dry pods

- Fodder types have mostly leafy growth. They produce only a few pods.
- Dual types produce both seed and fodder.

The seed

Seed types

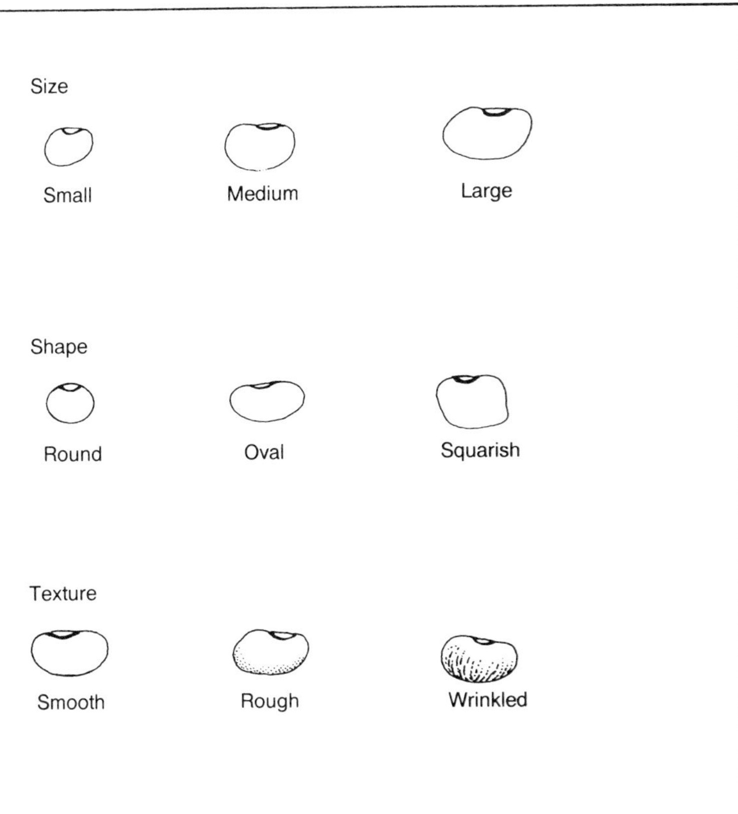

- Cowpea seeds vary in size, shape, color, and texture.
- Color may be white, black, red, or brown.

Parts of the seed

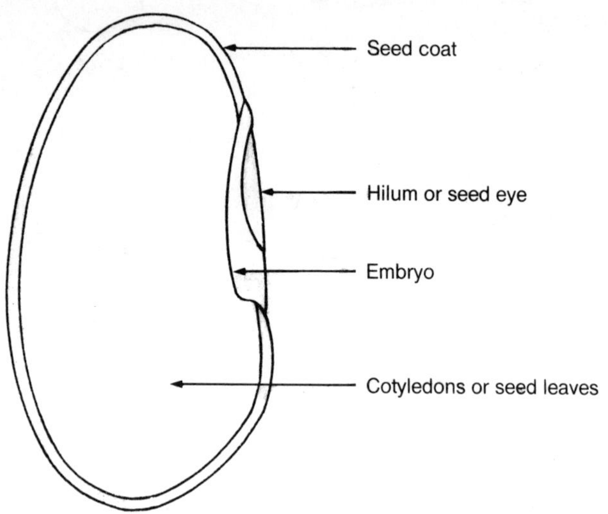

- Parts of the seed include the hilum, seed coat, cotyledons, and embryo.

Stages of germination

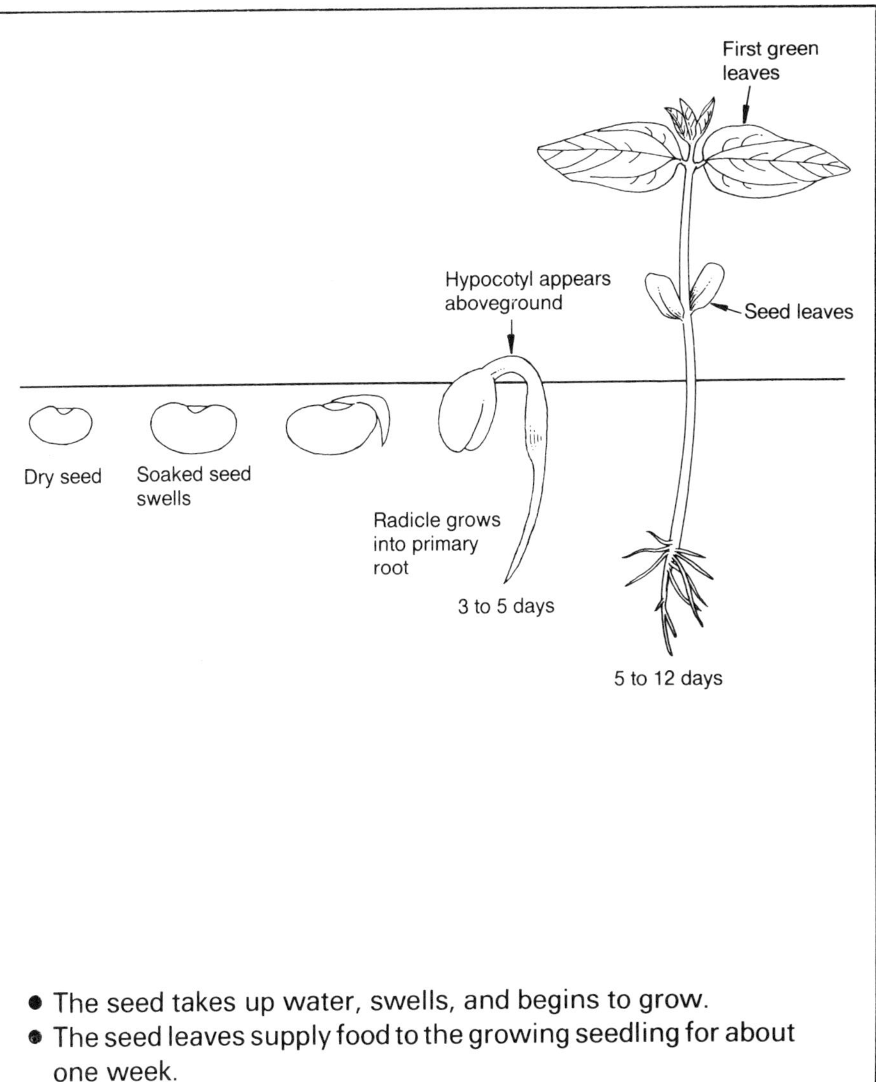

- The seed takes up water, swells, and begins to grow.
- The seed leaves supply food to the growing seedling for about one week.

Factors affecting germination — water, air, and warmth

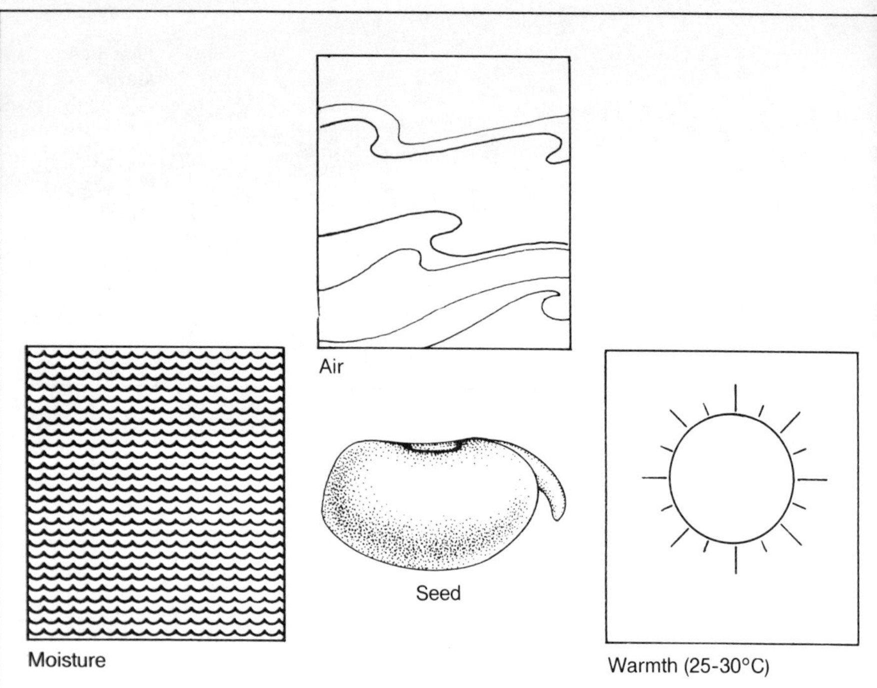

- To sprout and grow, the seed needs water, air, and warmth.
- With too little water, the seed will not start to grow. With too much water, it will rot.
- Without air the seed will mold or decay.
- Too much heat or cold will kill the growing embryo.

Factors affecting germination — seed quality

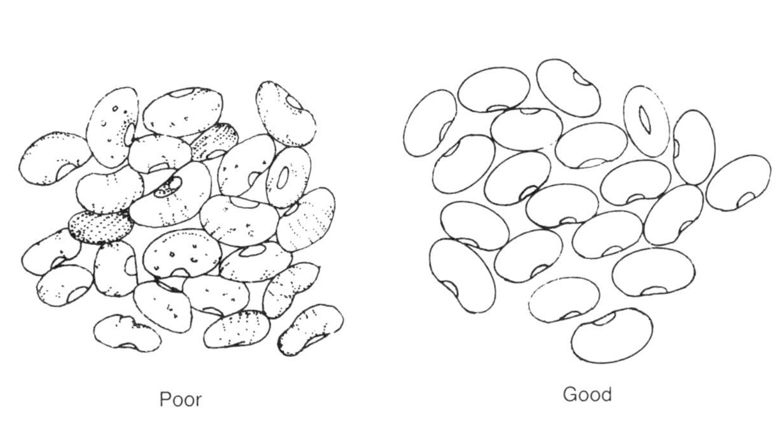

- For good germination, seed should be fresh, clean, and healthy.
- Treating seed with fungicide will help even germination.
- Seed for planting should be stored no more than 12 months.

Seedling growth

Factors affecting seedling growth — water

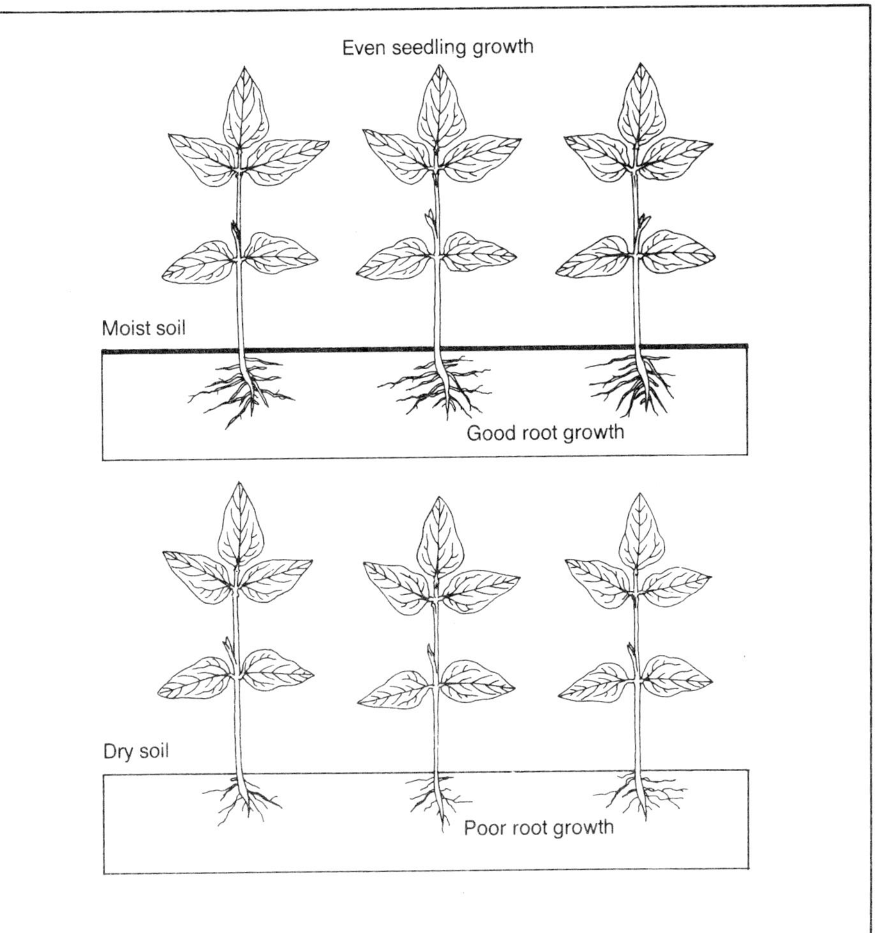

- Soil moisture is essential for even germination and seedling growth.
- Roots grow poorly in dry soil and cannot absorb nutrients for the plant.

Factors affecting seedling growth — temperature

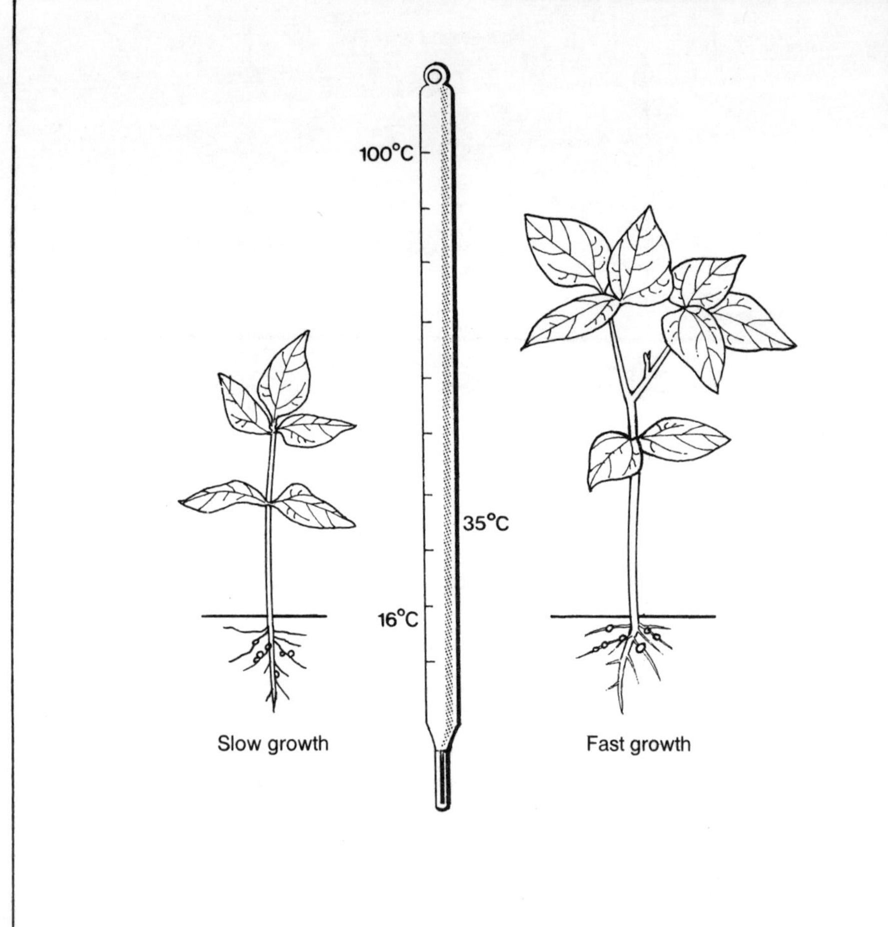

- Seedlings grow fast in warm weather. Cold weather slows down growth and seedlings cannot compete with weeds.

Factors affecting seedling growth — light

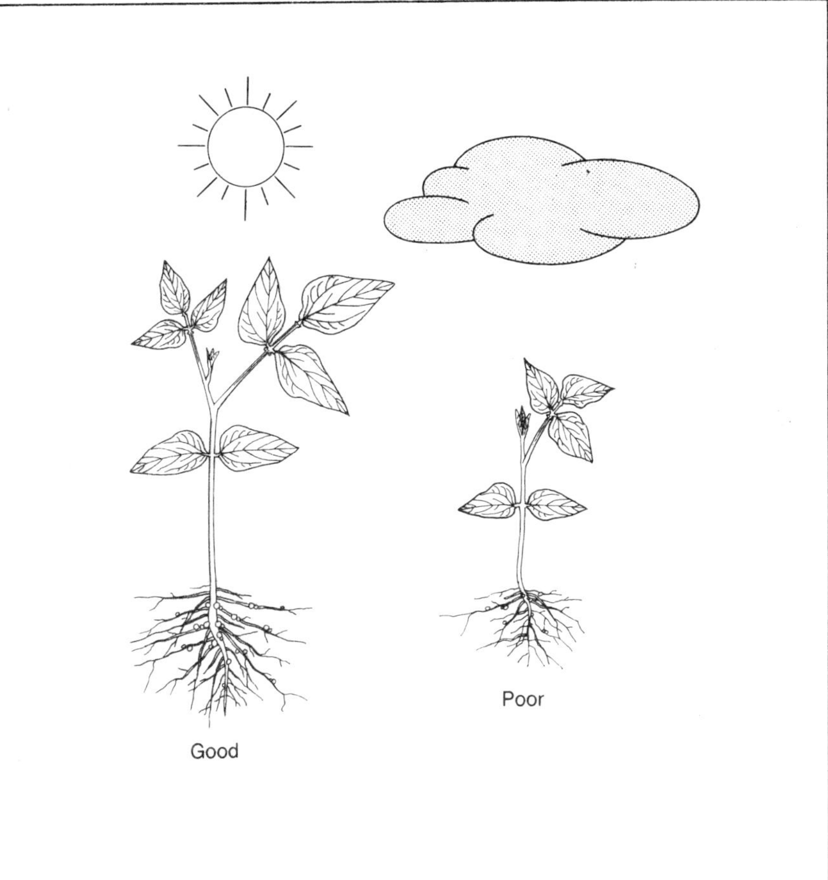

- Bright sunlight helps vigorous seedling growth. Plant cowpea in sunny areas, away from shade trees.

Factors affecting seedling growth — nutrients

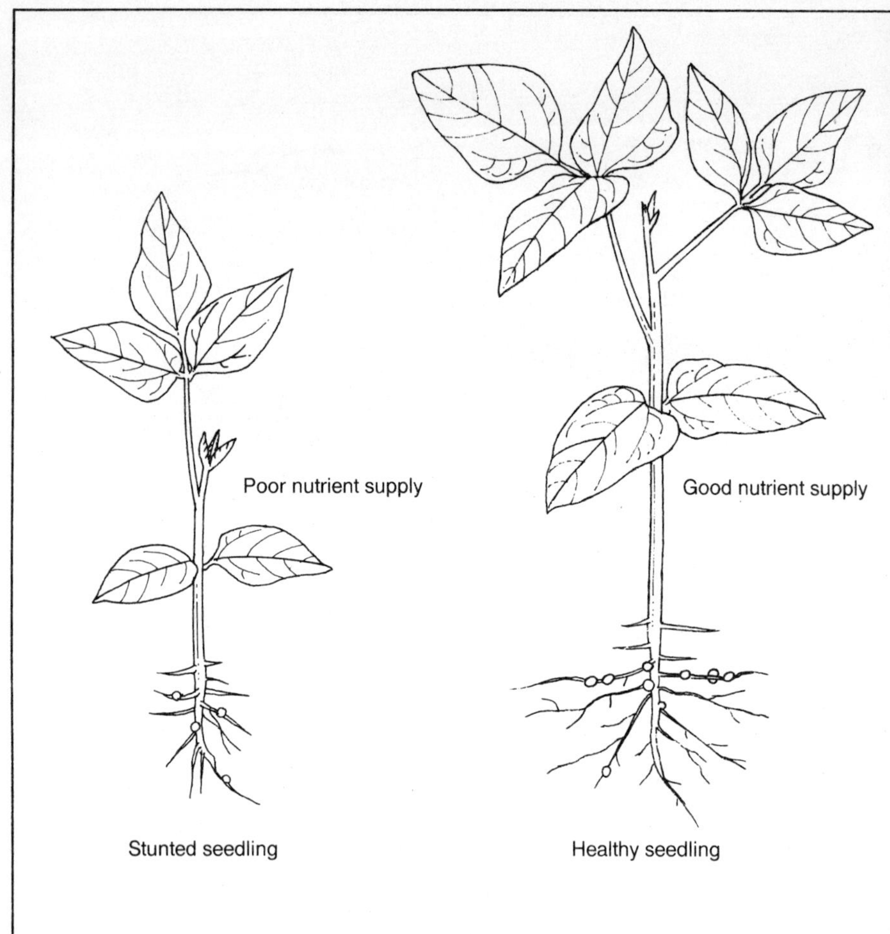

- Usually cowpea can grow on nutrients left over in the soil from the rice crop. But in poor soils, fertilizer added at planting starts rapid growth.

Factors affecting seedling growth — plant density

- Seedlings growing too close together grow too tall and lodge easily.
- Seedlings spaced too far apart allow too much weed growth.

Factors affecting seedling growth — weeds and insects

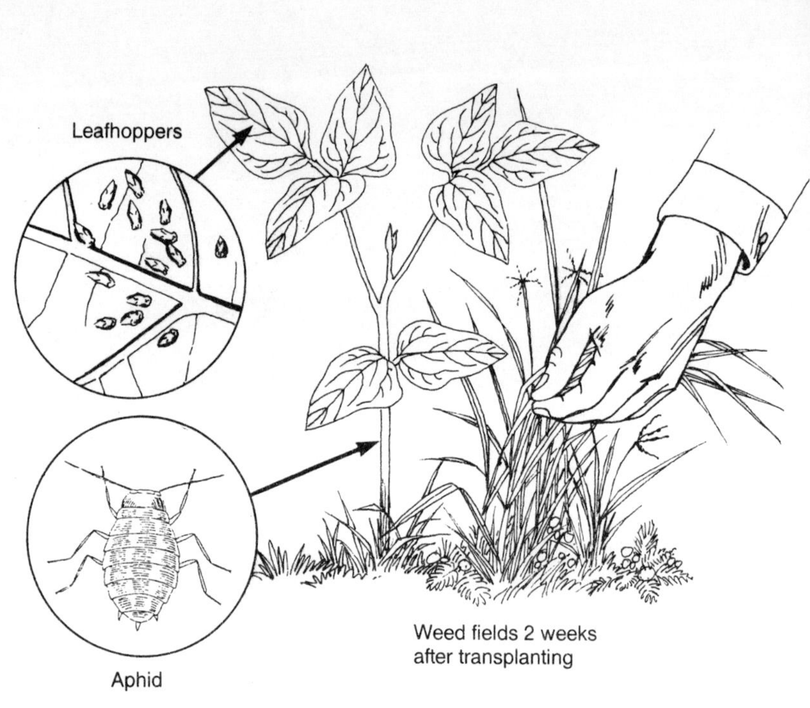

- Weeds rob seedlings of nutrients.
- Insect pests that eat young leaves and stems may kill seedlings.

Growth stages

Growth stages

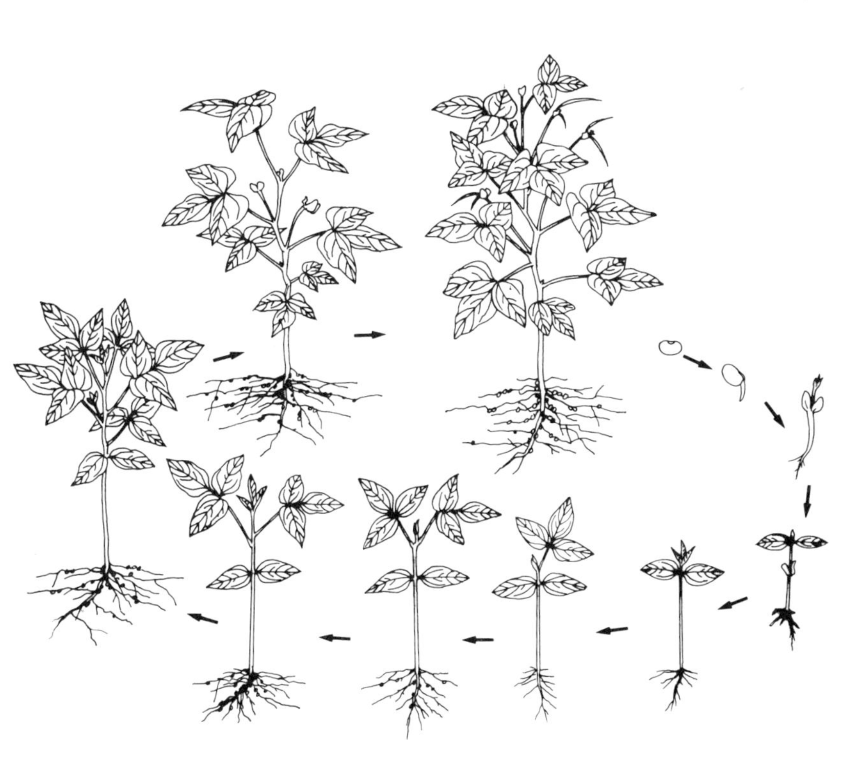

- The cowpea plant goes through eleven growth stages from
  seed germination to maturity.

Growth stages

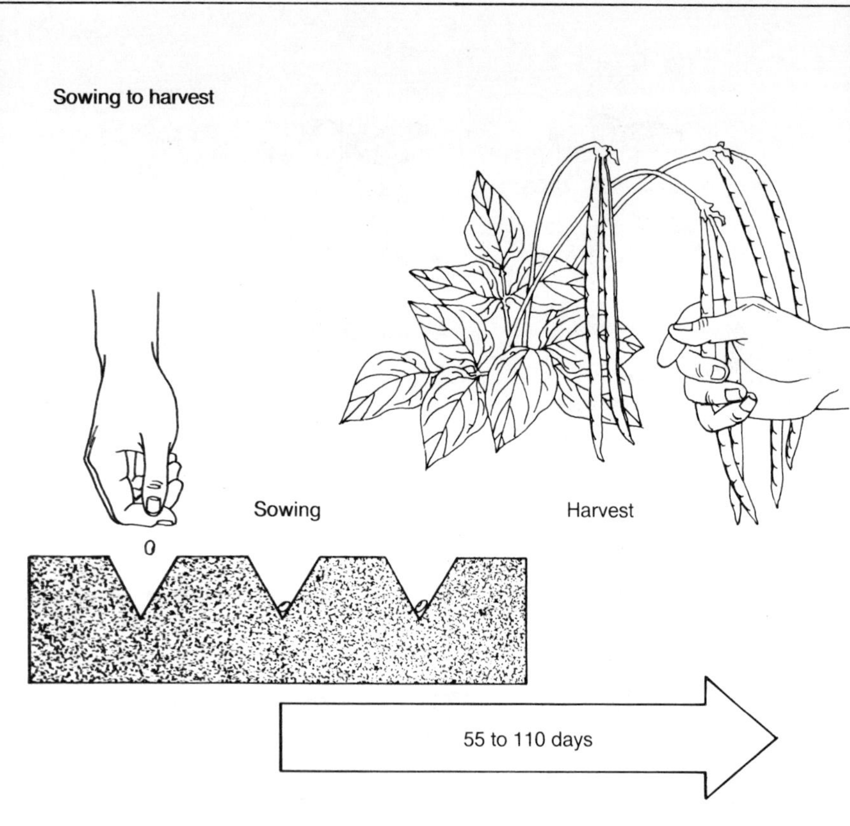

- Sowing to harvest may take 55 to 110 days, depending on variety, season, and growing conditions.

Vegetative phase

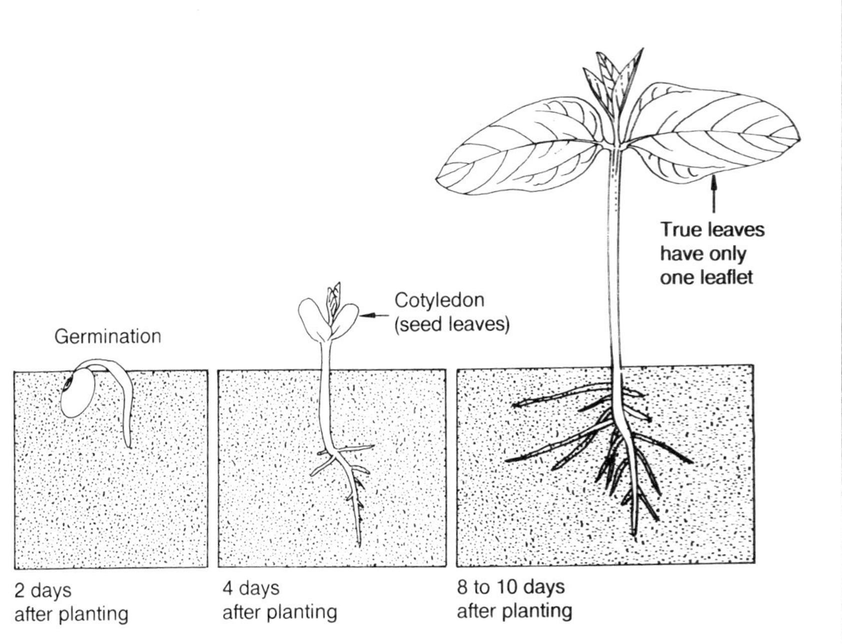

- The vegetative phase lasts from germination until the first flower appears, about 40 days after planting.
- The first pair of true leaves unfolds on the ninth to eleventh day after planting.

Vegetative phase

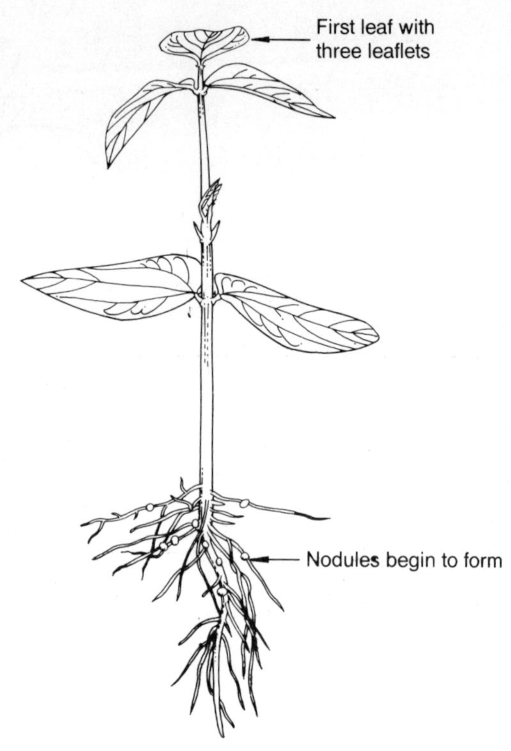

- At 13 to 15 days after planting, the first leaf with three leaflets unrolls.
- Nodules begin to form on the roots.

Vegetative phase

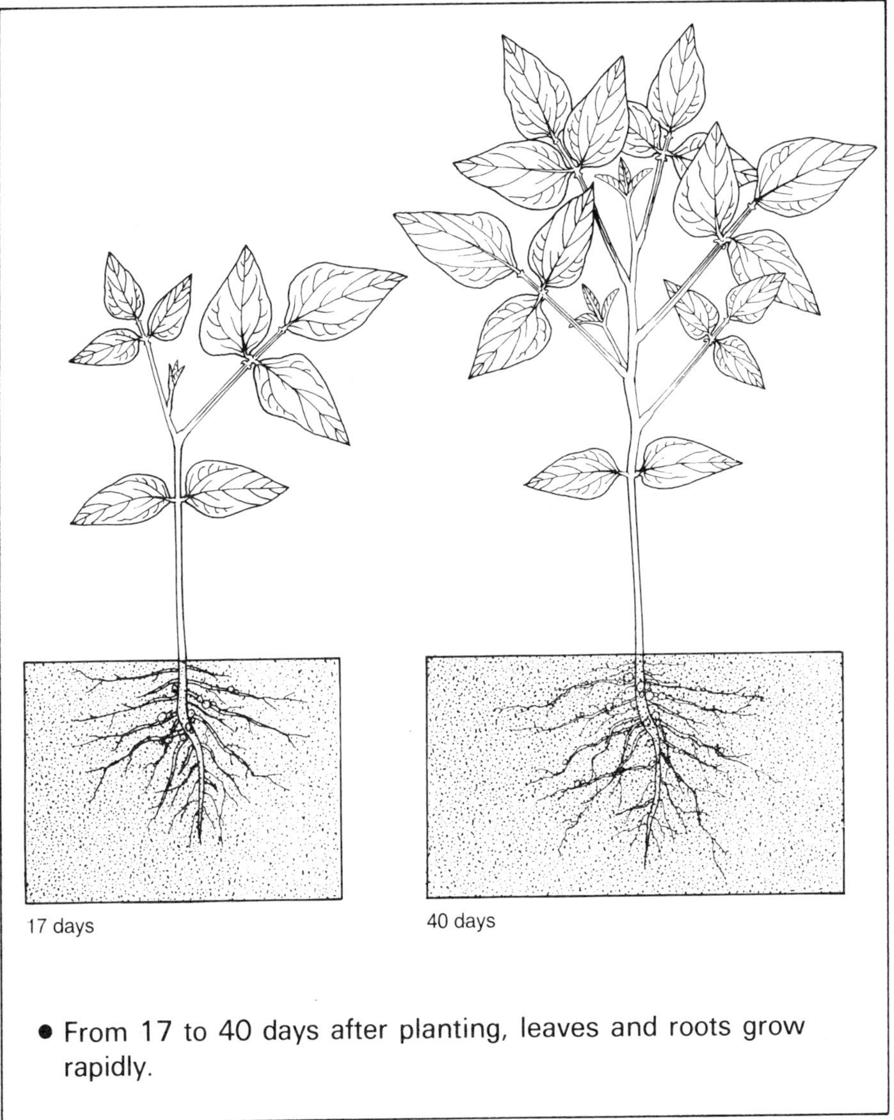

- From 17 to 40 days after planting, leaves and roots grow rapidly.

Vegetative phase

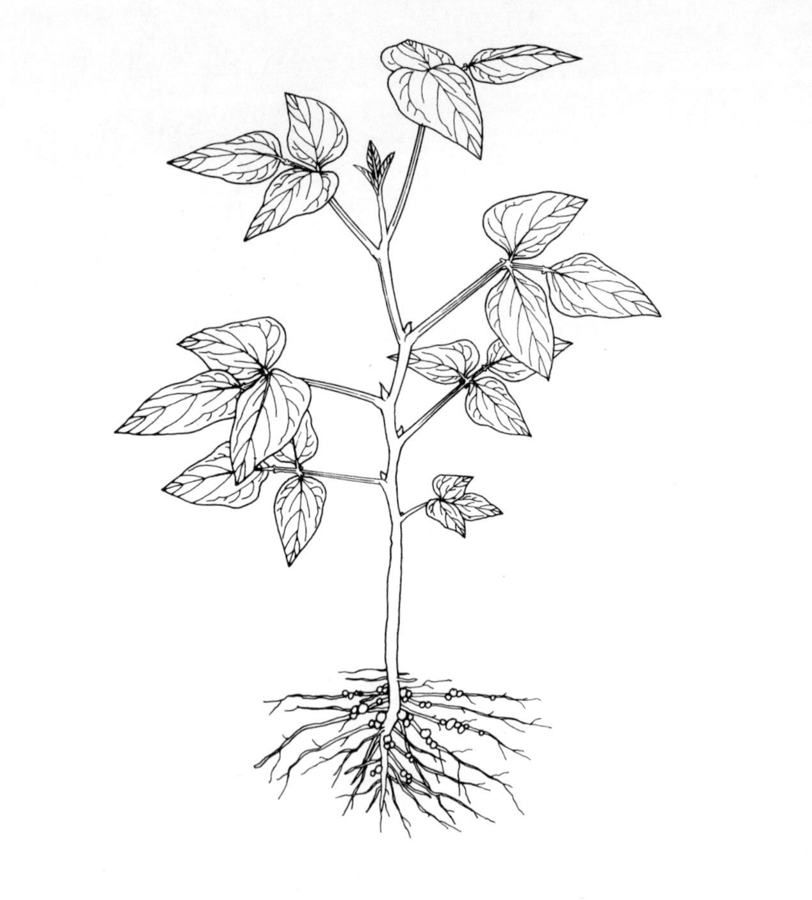

- Root nodules develop to the maximum and the plant fixes nitrogen at a high rate.

Reproductive phase — flowering

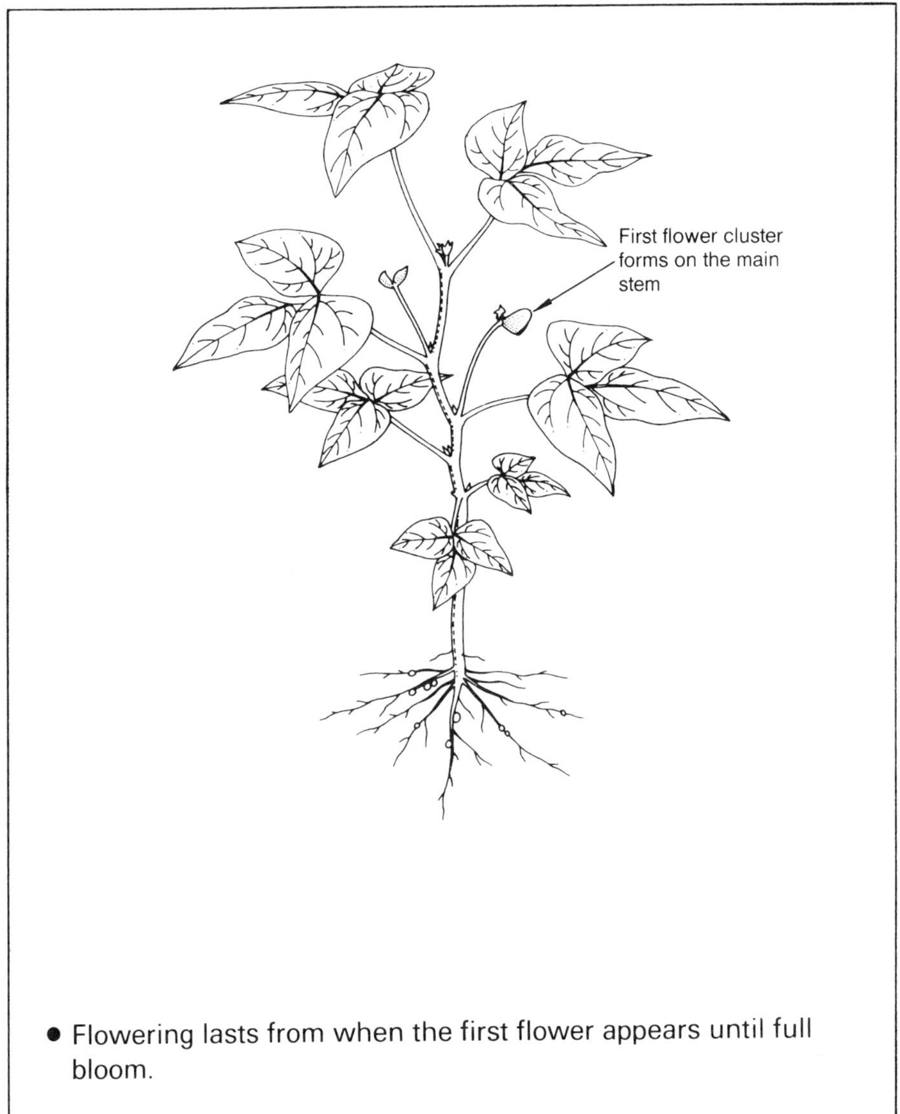

- Flowering lasts from when the first flower appears until full bloom.

Reproductive phase — pod formation

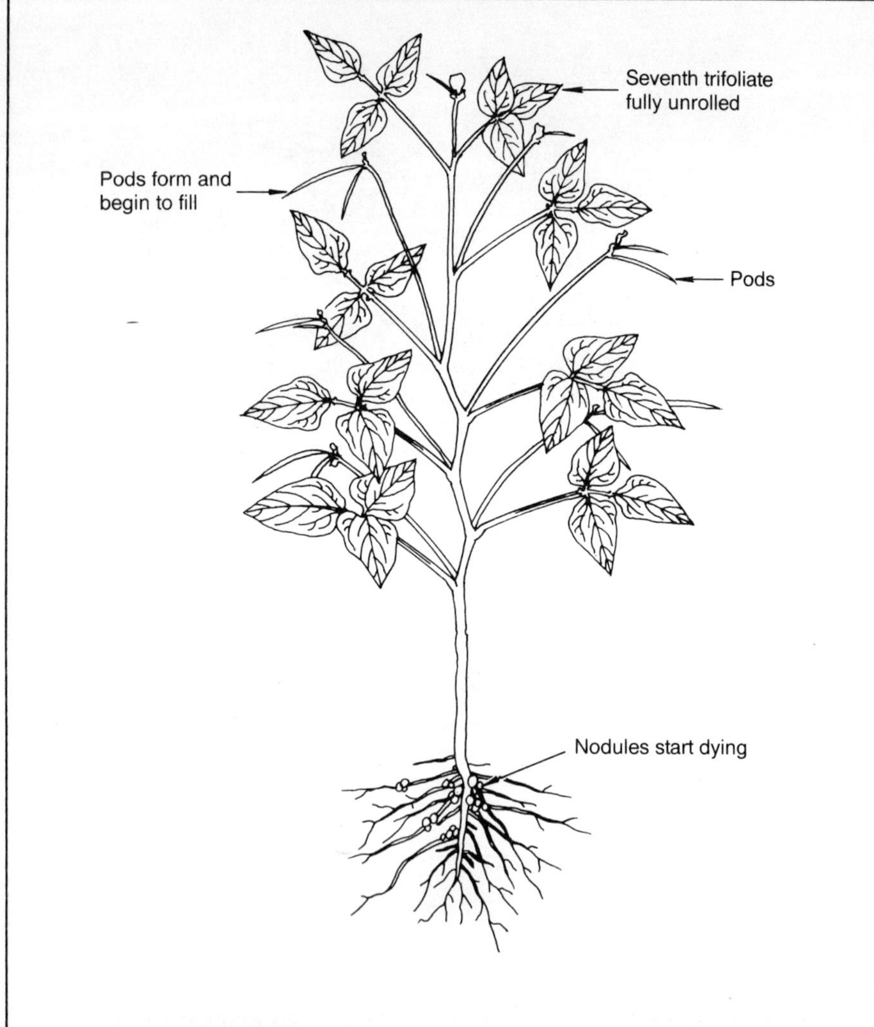

● Pods form and begin filling after the seventh leaf unrolls.

Reproductive phase — ripening and maturity

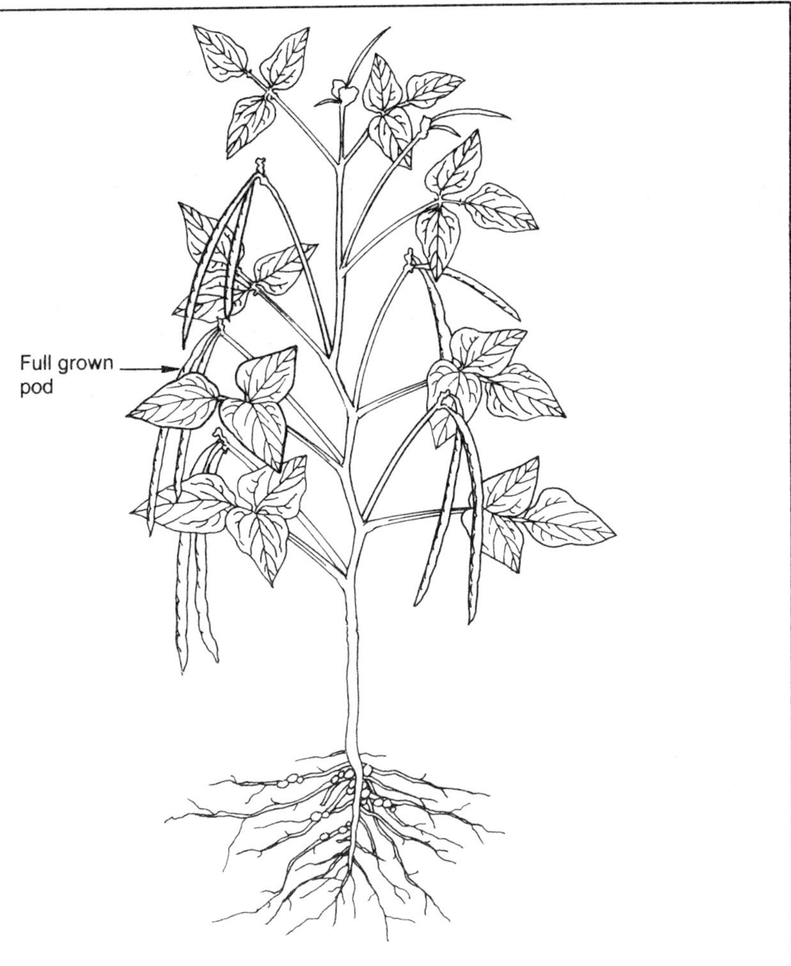

- Fully developed pods are dark green.
- As they ripen and mature, they change to brown, purple, or gray.

The roots

Origin of roots

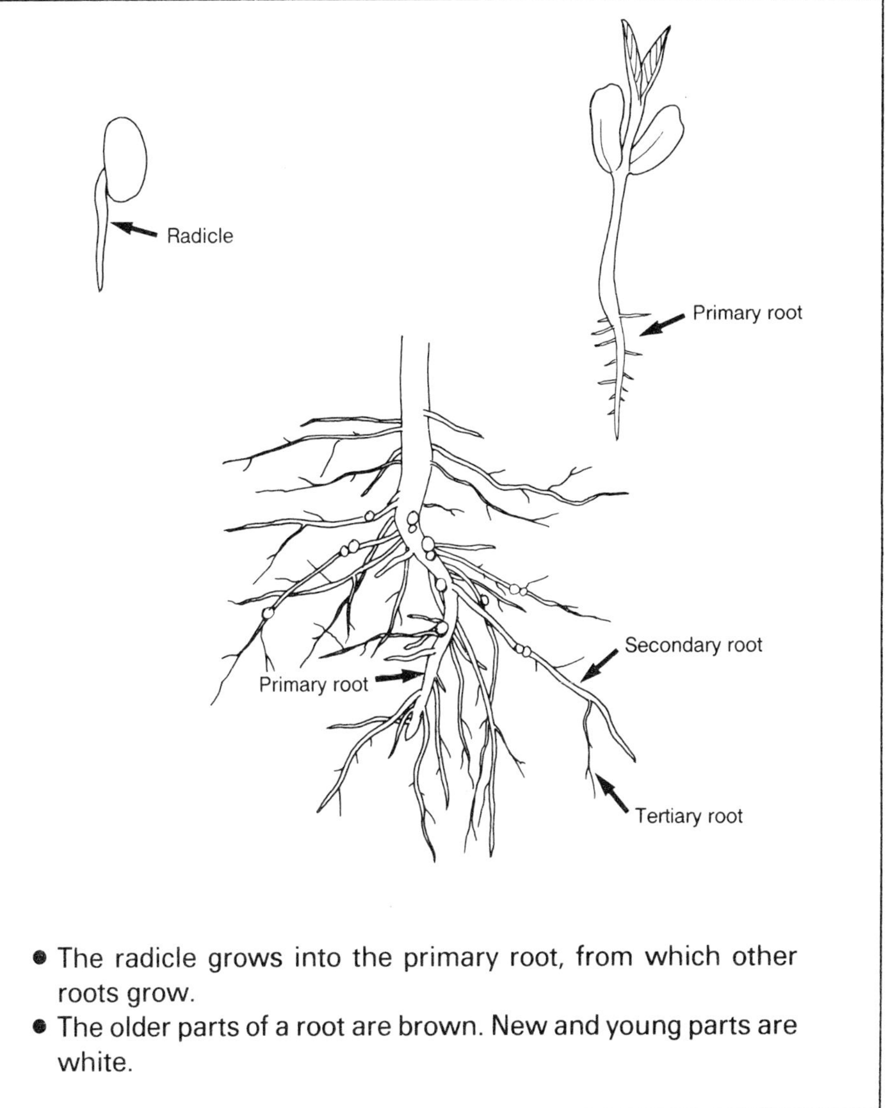

- The radicle grows into the primary root, from which other roots grow.
- The older parts of a root are brown. New and young parts are white.

Functions of roots

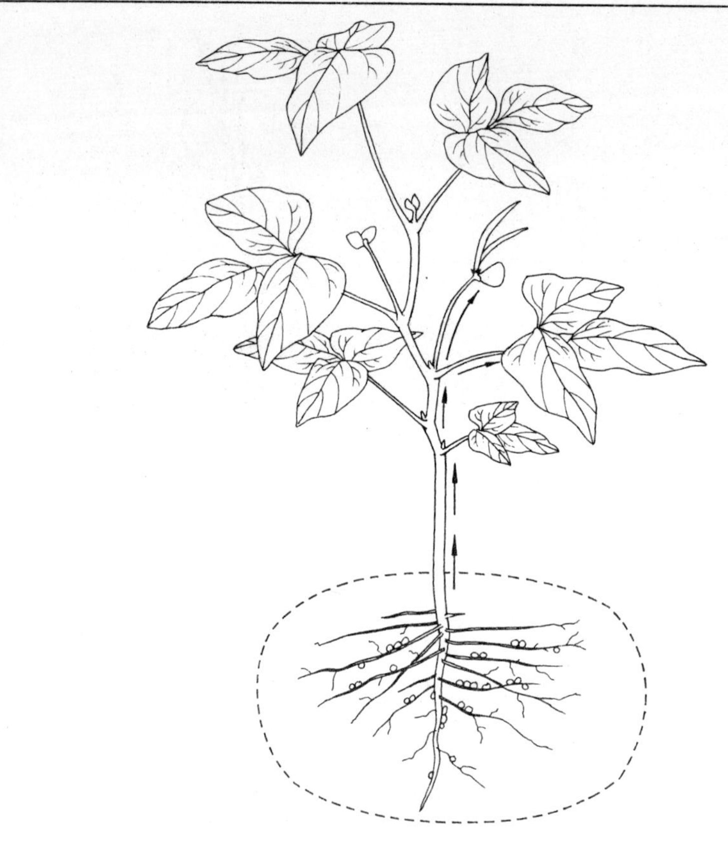

- Roots transport water and nutrients to leaves, flowers, and pods.
- They support the upper parts of the plant.
- Roots in cowpea are also sites of nitrogen fixation.

Root distribution

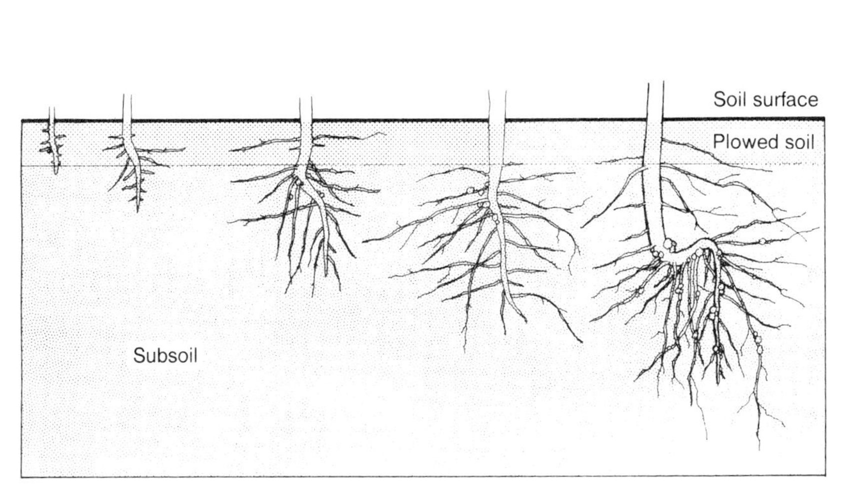

- The roots grow rapidly as soil water dries out.
- Most of the roots remain in the upper soil layer. Only a few go down into the lower layer.

Root development — emergence to flowering

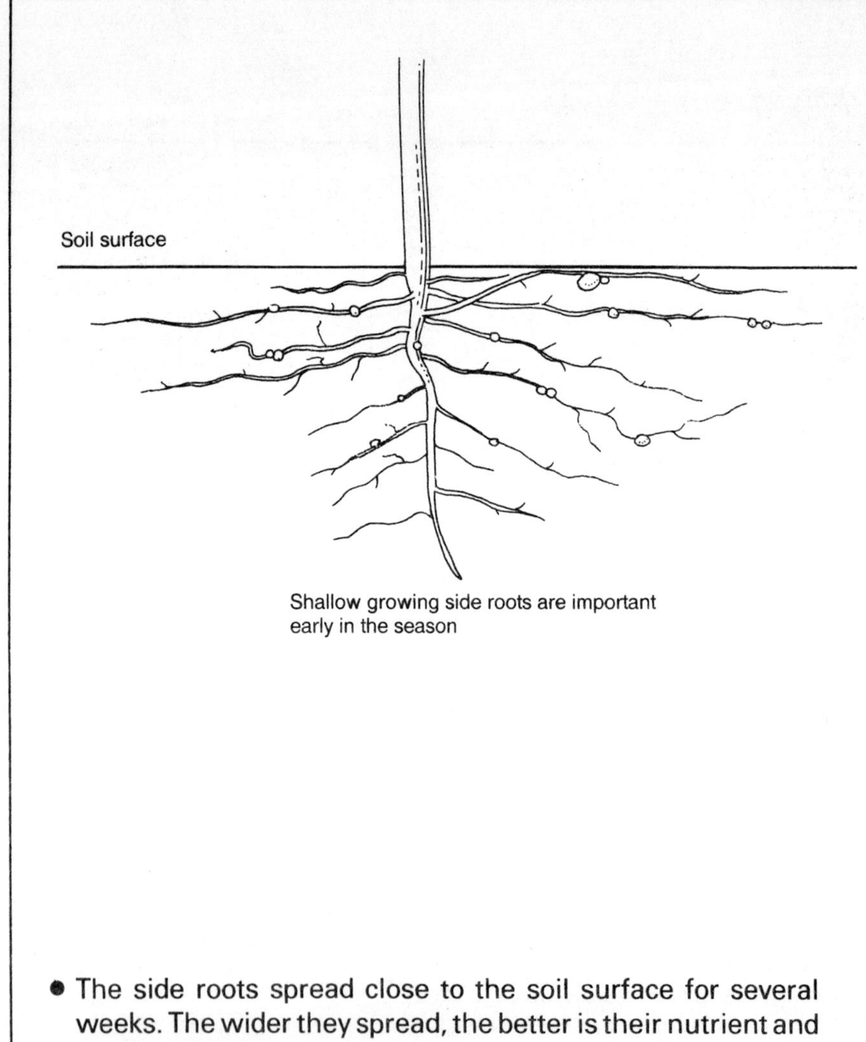

- The side roots spread close to the soil surface for several weeks. The wider they spread, the better is their nutrient and water uptake early in the season.

Root development — flowering to pod ripening

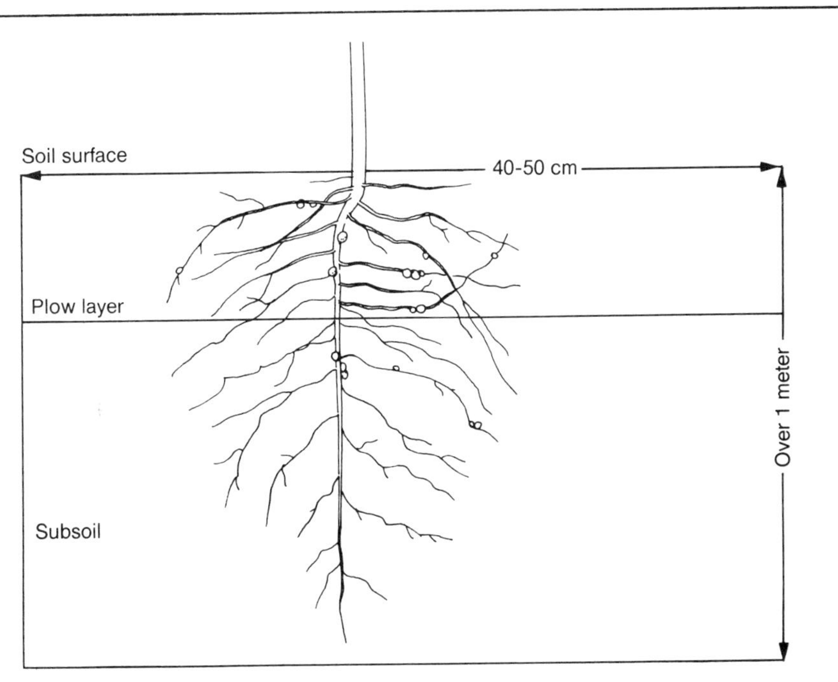

- The lower roots grow deep into the subsoil as soil water dries out.
- The deeper they grow, the more water they can absorb for crop growth and yield.

Root nodules and nitrogen fixing

Root nodules

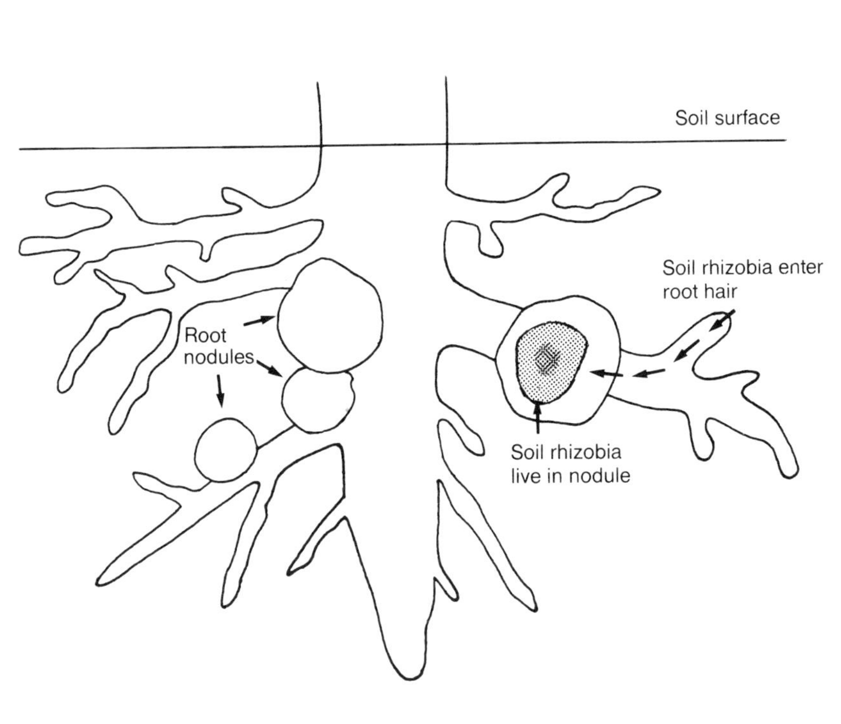

- Nodules are small lumps that grow on cowpea roots.
- Soil bacteria called rhizobia live in these nodules and fix nitrogen from the air, which the plant uses.

Root nodules

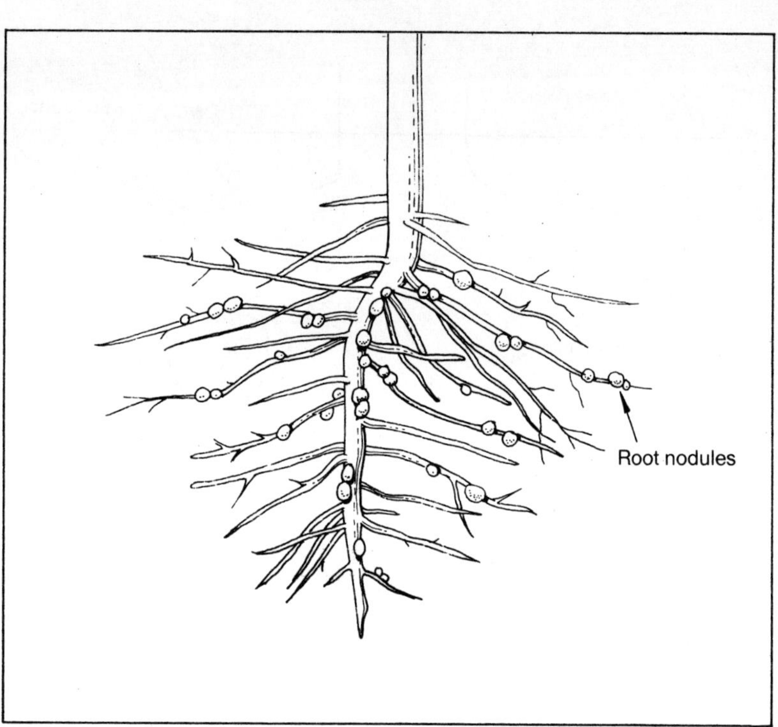

- Healthy nodules are important to good crop growth.
- Nodules appear on the roots about 15 days after seedlings emerge. Nodulation reaches a peak during flowering and early pod formation.

Nitrogen fixing

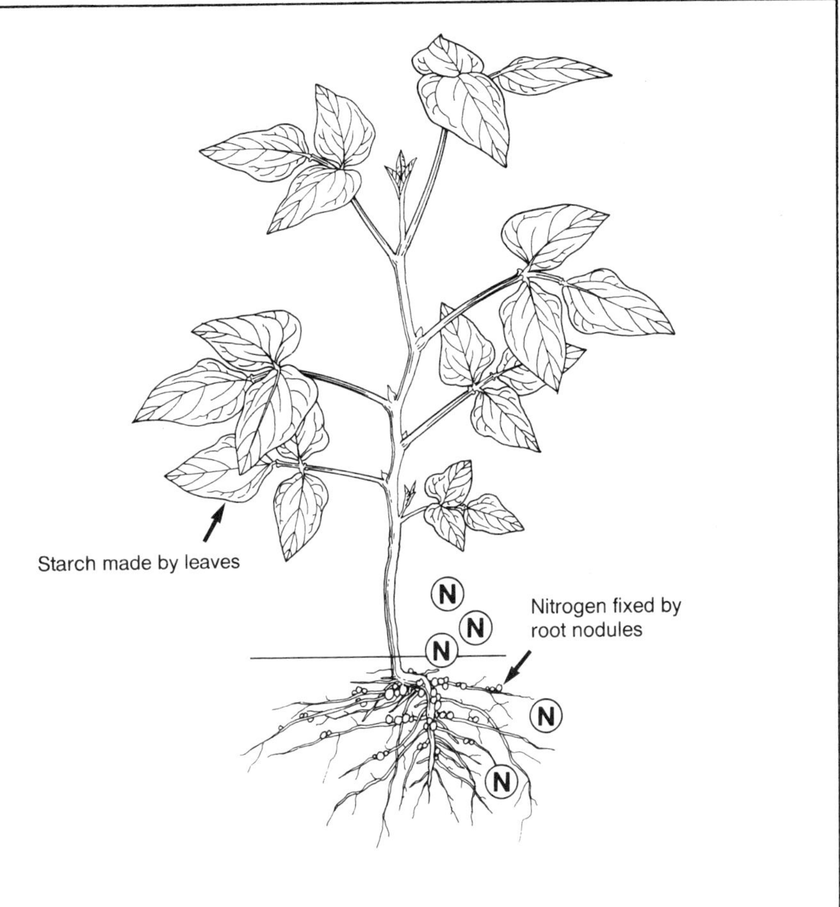

- Nitrogen fixing begins soon after nodules form. The fixing rate is highest during flowering and early pod formation.
- After this nodules begin to die and nitrogen fixing decreases.

Conditions affecting nitrogen fixing — soil nitrogen and phosphorus

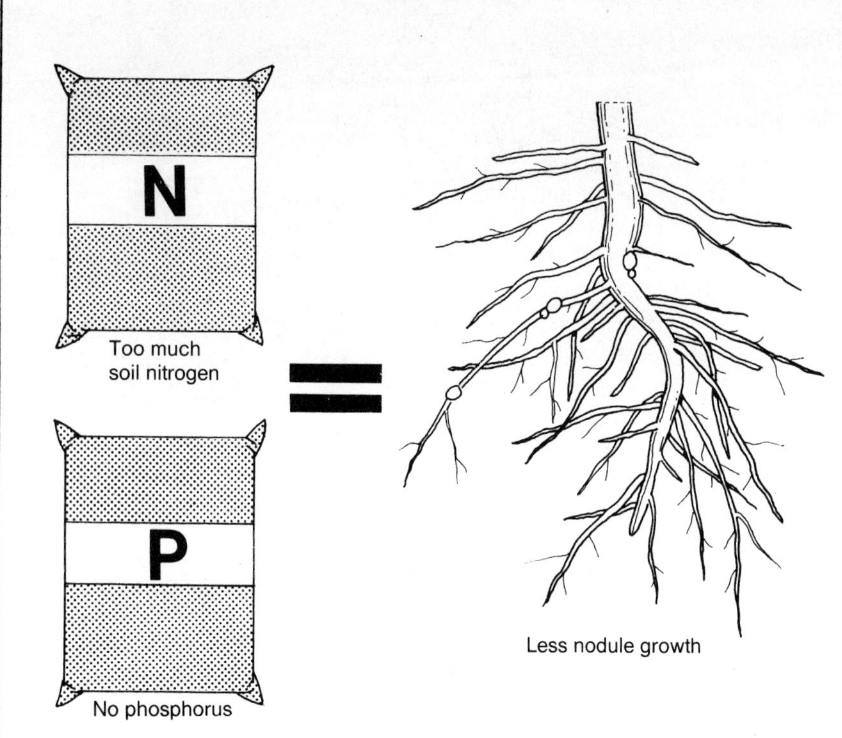

- Too much soil nitrogen reduces nodule growth and activity.
- Lack of phosphorus also reduces nodule growth.

Conditions affecting nitrogen fixing — temperature and daylength

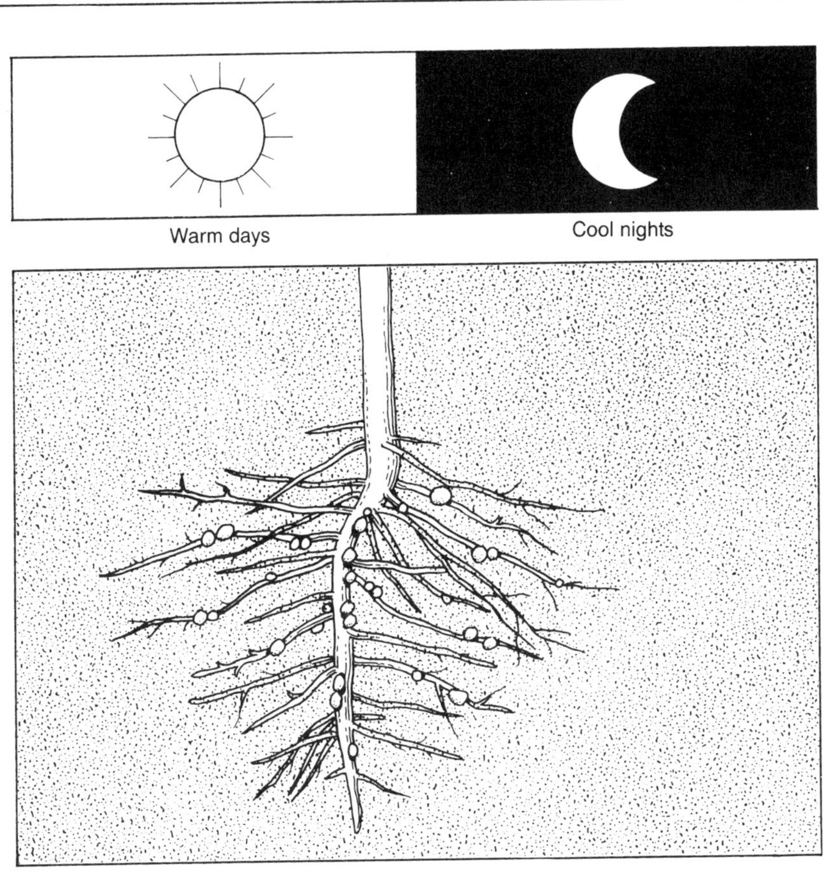

- Warm days and cool nights increase nodule activity.
- Daylength should be less than 16 hours.

Conditions affecting nitrogen fixing — soil rhizobia

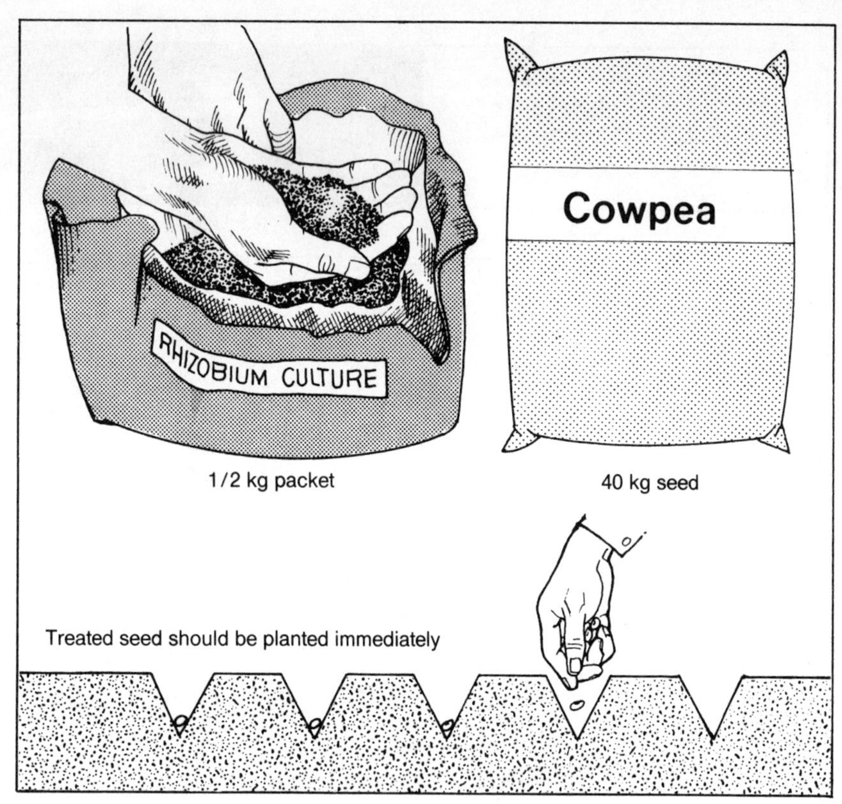

- In fields where legumes have not been grown for more than 5 years, cowpea seed must be treated with Rhizobium culture before planting.
- Culture is available in packets at farm supply centers or from extension agencies.

The shoot — leaves and branches

The cowpea leaf

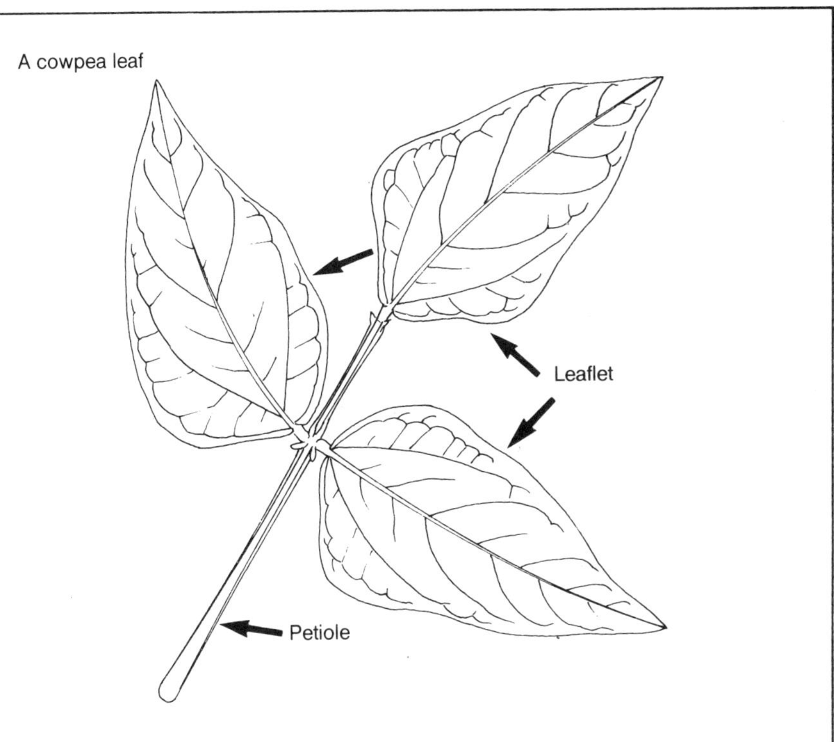

- The green leaves trap sunlight to manufacture food for the plant, using water from the soil and carbon dioxide from the air.

Canopy development

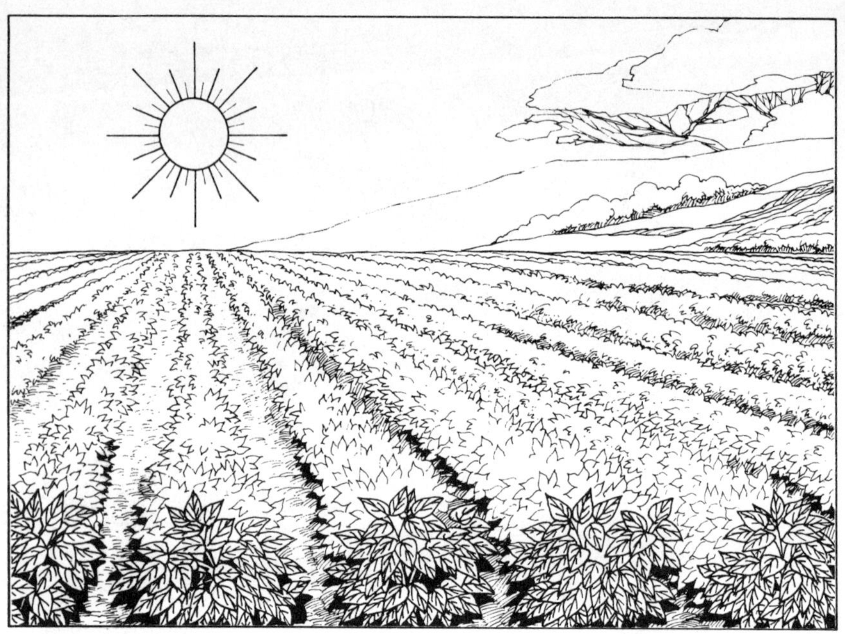

A good crop canopy
- Traps sunlight
- Shades the soil and keeps it moist
- Reduces weed growth

- In a healthy cowpea crop, the upper leaves form an umbrella, or canopy, shading the ground between rows.
- Some sunlight should get through to the lower leaves.

Loss of leaves

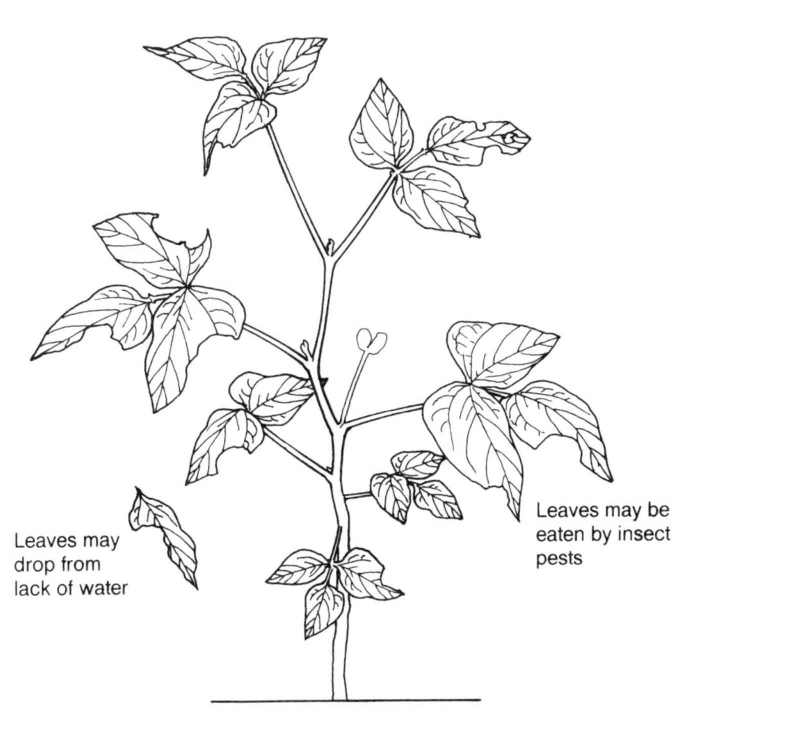

- Loss of leaves from lack of water or insect damage means less carbohydrate to nourish the plant.
- The plant will produce fewer flowers and pods.

Branches

- Branching starts 2 to 3 weeks after emergence.
- Branches are useful in making up some yield where plant numbers are low. But they cannot make up for poor plant stands.

The shoot — flowers and seed pods

Flowering

- The first flower stalk develops from the middle of the plant, in the axil between leaf and stem.
- Flowering progresses upwards and downwards from here.

Pod formation

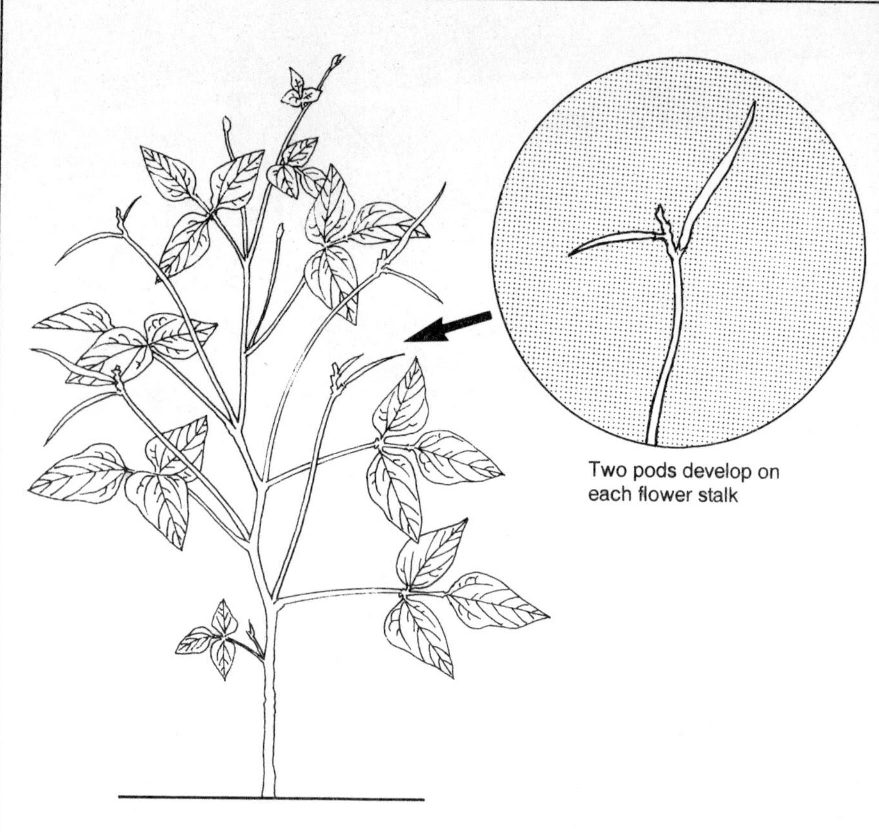

- A pod begins to form when the male cell from the pollen unites with the egg in the ovary.
- Usually only two flowers on each stalk develop into pods.

Flower and pod drop

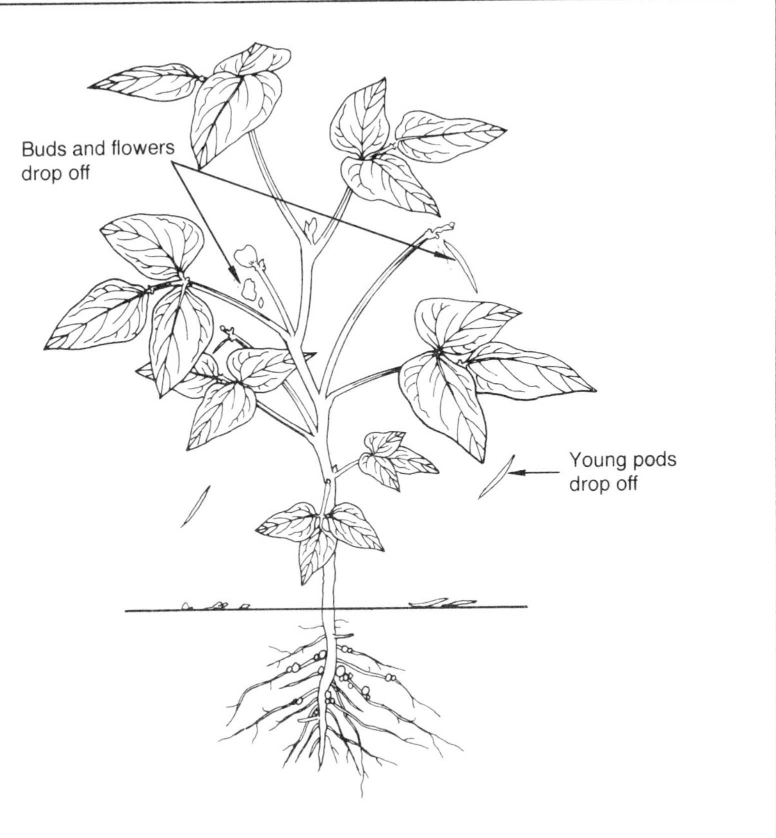

- Fifty to sixty percent of the buds and flowers drop off the plant. Sometimes young pods also drop.
- Proper water and nutrient supply at flowering and pod filling will reduce flower and pod drop and increase number of mature pods.

Stages of pod filling

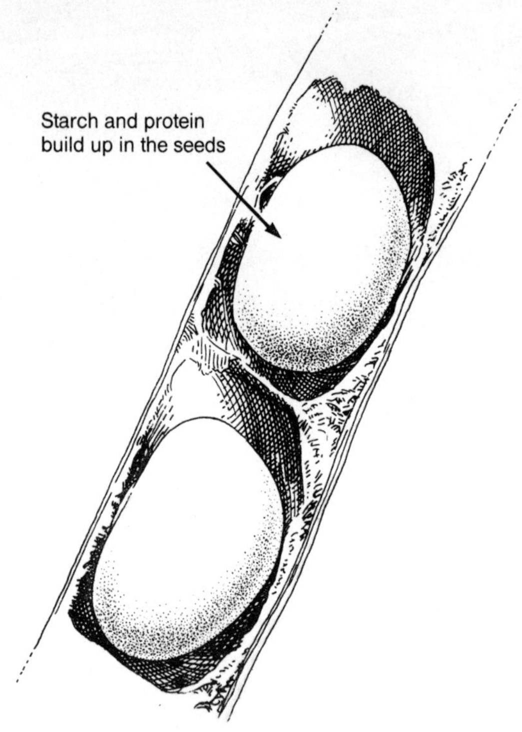

- Starch and protein builds up in the seeds. The pod wall thickens and becomes tough as the pod develops.

Pod filling

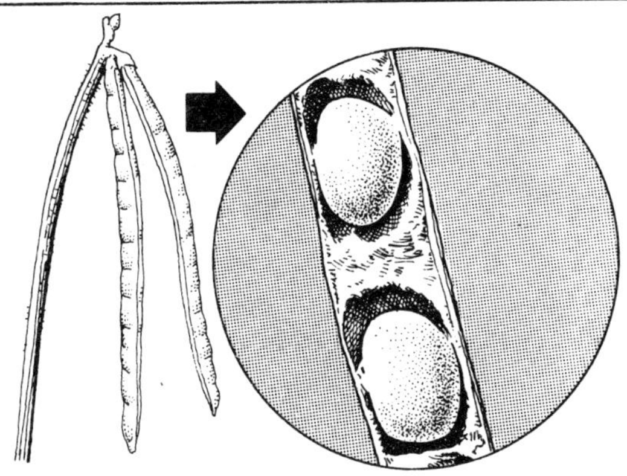

Fully filled pod

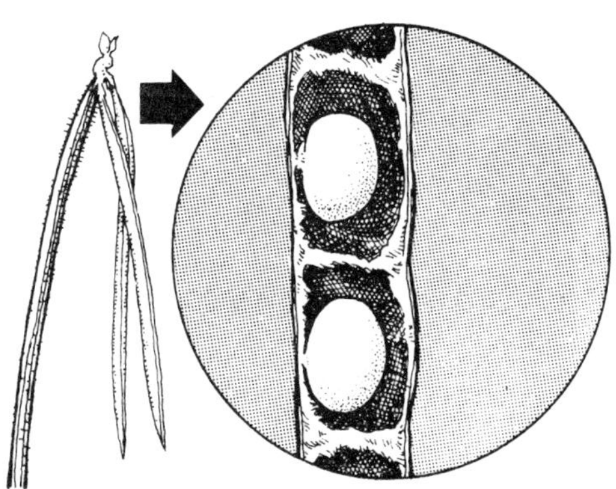

Pod with some unfilled seed

- Seeds develop over 20 to 25 days. They fill slowly for the first few days and then rapidly.

Dry matter production

Dry matter production

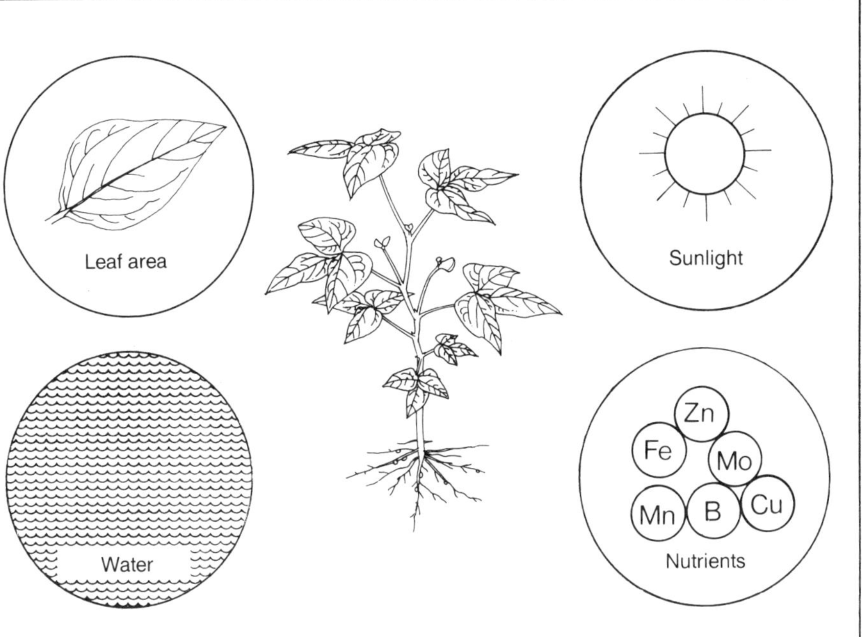

- Fresh plant weight minus water gives total dry matter in a crop.
- Dry matter accumulation is important to the total yield of both seed and fodder.
- Cowpea plant dry matter contains mostly starch, fiber, and protein.

Factors affecting dry matter production — leaf area

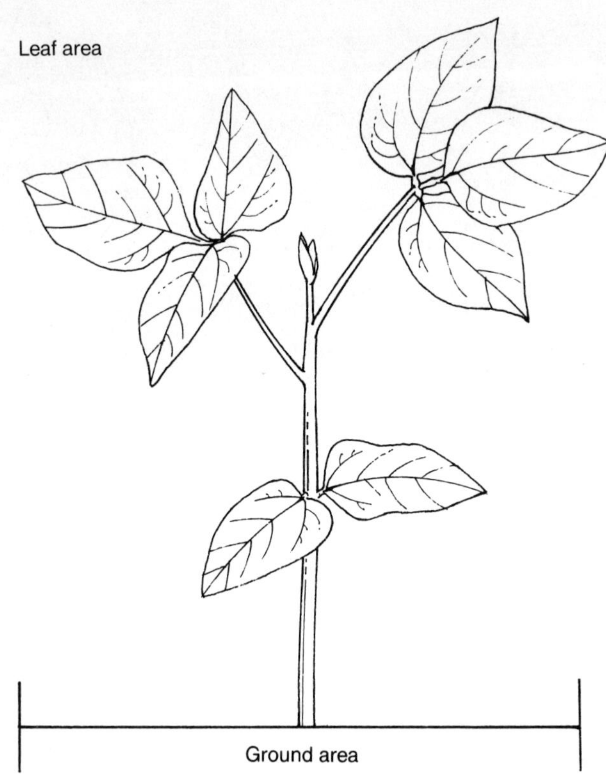

- Leaf area depends on number of plants per square meter, and on water and nutrient supply.
- A high leaf area index will give higher dry matter production.

Factors affecting dry matter production — sunlight

Shade reduces dry matter accumulated

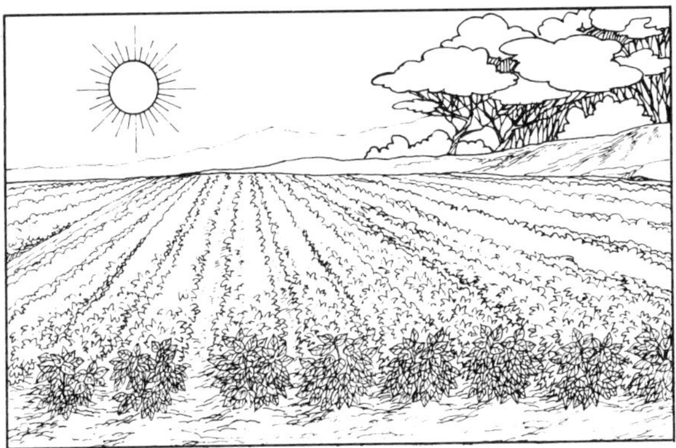

Light increases dry matter produced

- Bright sunlight increases dry matter produced.
- When cowpea is grown in shade, as on a coconut plantation, dry matter will be reduced as shade increases.

Factors affecting dry matter production — water

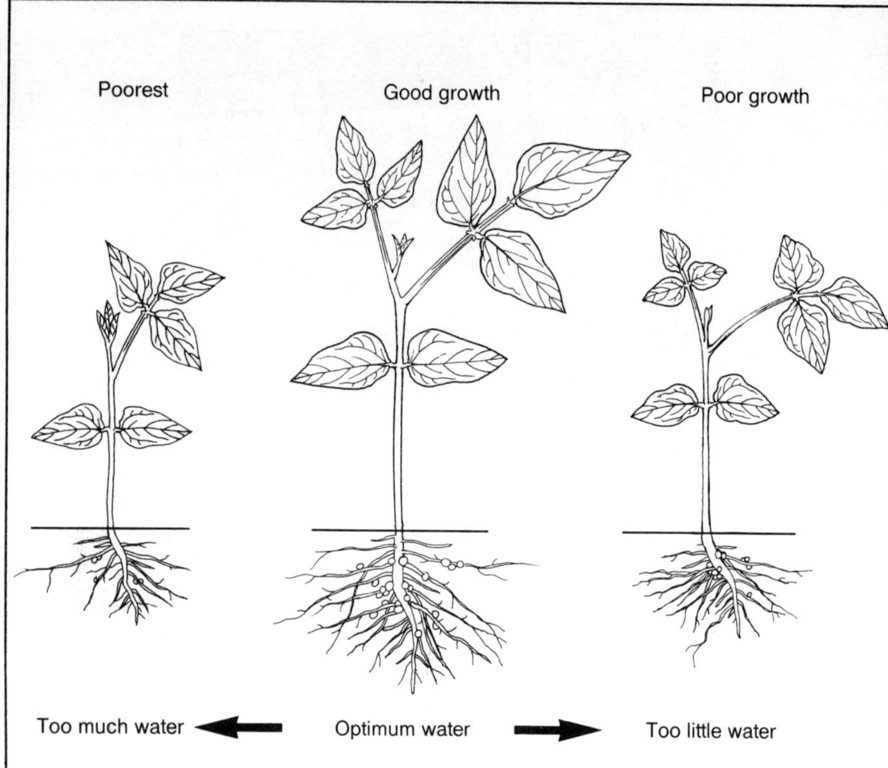

- Maximum dry matter is produced when the plant gets the right amount of moisture.
- With too little water, the leaf pores close, reducing food made by the leaves.
- If soil is waterlogged the roots cannot absorb nutrients.

Factors affecting dry matter production — nutrients

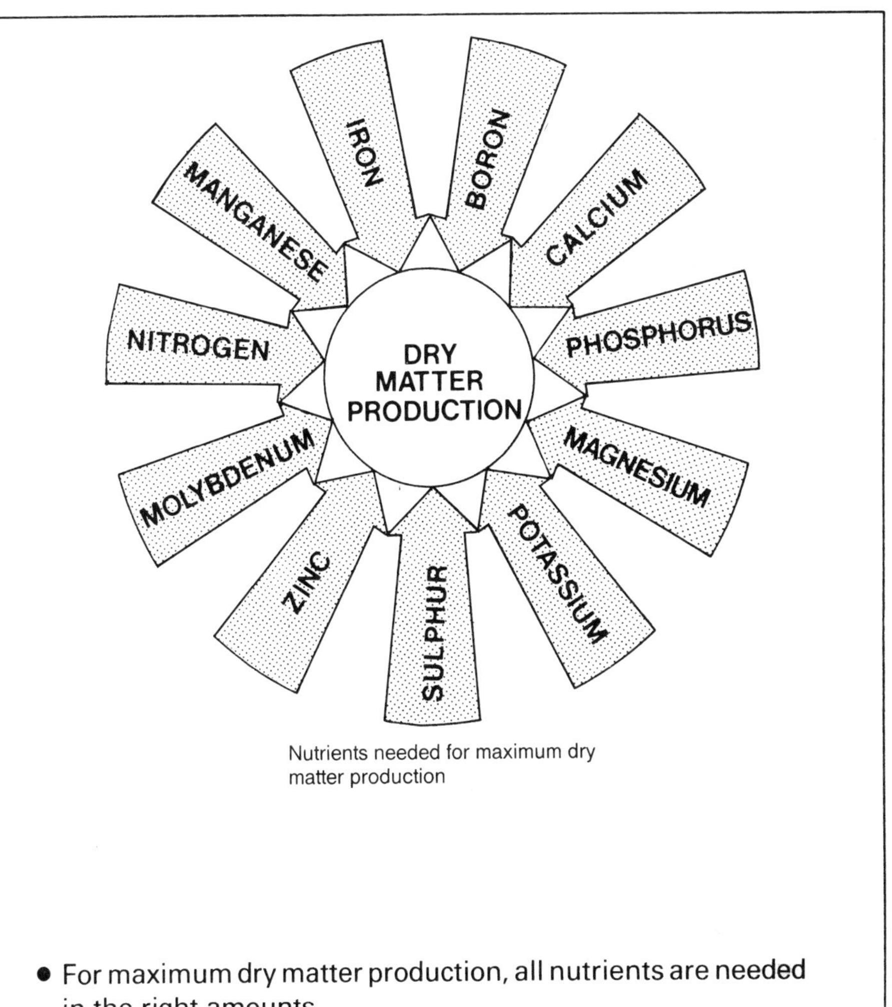

Nutrients needed for maximum dry matter production

- For maximum dry matter production, all nutrients are needed in the right amounts.
- Lack of any nutrient will sharply reduce dry matter, even if the other nutrients are well supplied.

Growing cowpea

Growing cowpea — environment

Temperature

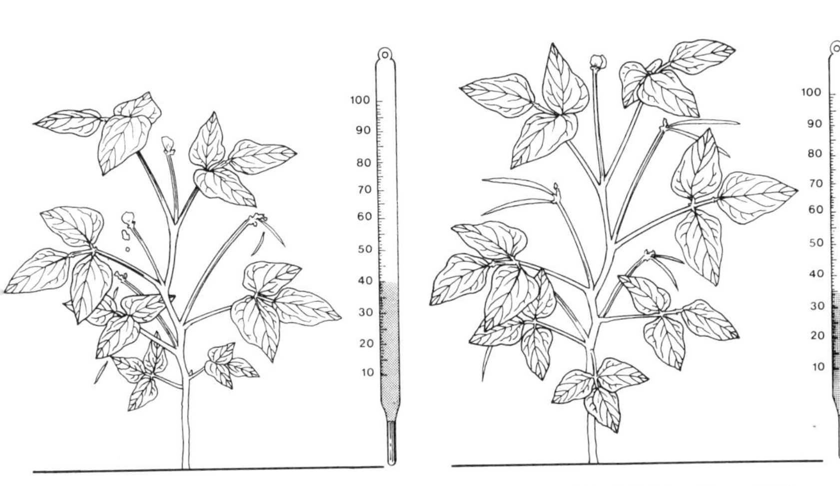

Above 38°C, flowers and pods may drop

20 to 35°C best for growth

- Cowpea is a tropical crop suited to hot, humid climates and semi-dry areas.
- The best temperature for growth is 20 to 35°C.
- Cowpea can stand low temperatures, down to 15°C, but not frost.

Rainfall

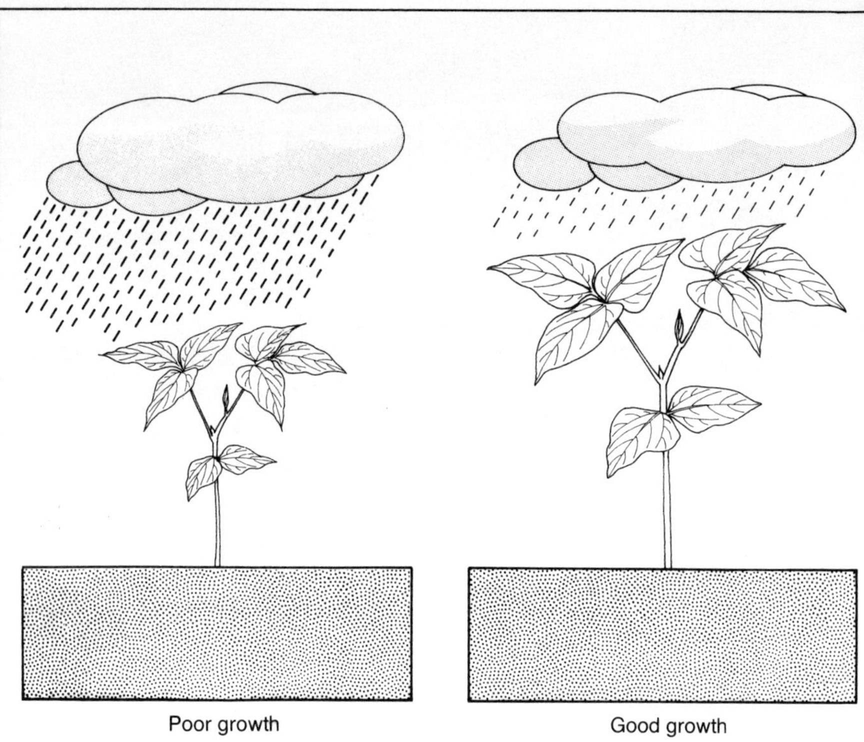

- Cowpea can grow in both low and high rainfall areas. But standing water will kill the plants.
- Drought during early growth stages will reduce yields.
- Rain during pod ripening gives poor seed quality.

Daylength

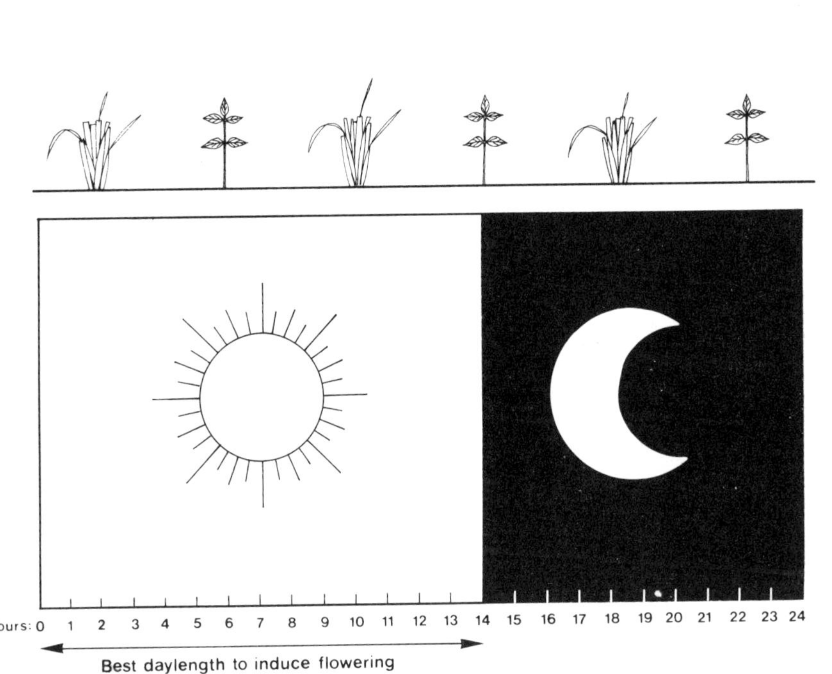

- Flowering is best when days are 8 to 14 hours long.
- Varieties that grow well under any daylength can be used throughout the tropics.

Light intensity

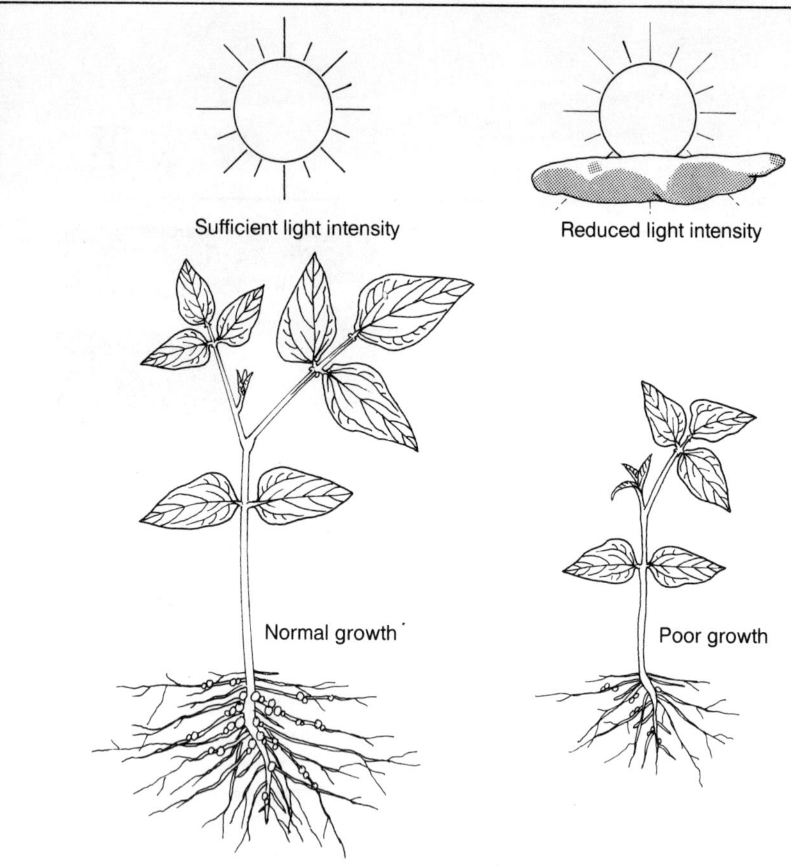

- Most varieties grow poorly in shade or reduced light. Leaves turn pale and stems are weak.
- Shade-tolerant varieties are available for growing with plantation crops such as coconut, oil palm, and rubber.

Soil

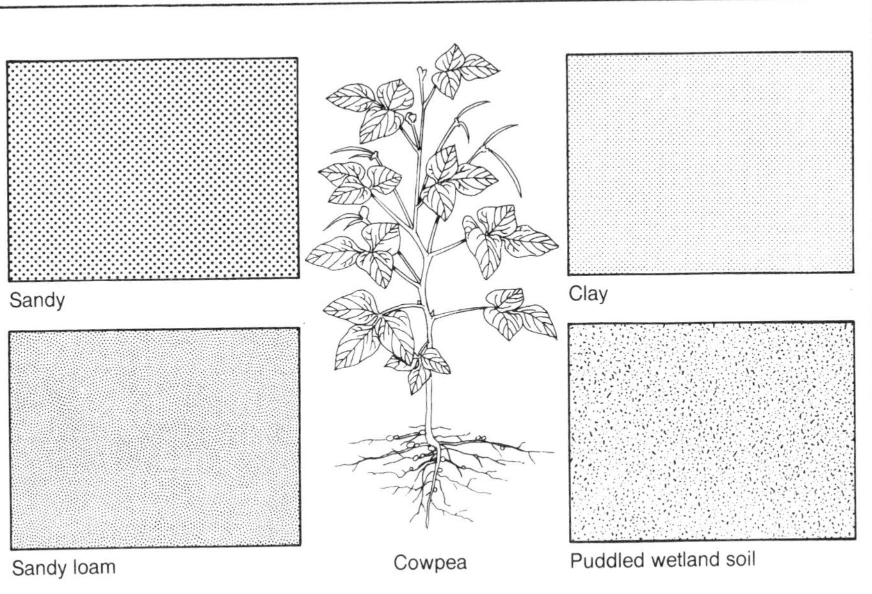

- Cowpea can grow on many kinds of soil, from sandy soil to clayey black soil.
- It can grow on puddled wetland rice soils and even on acid soils where mungbean and soybean cannot grow.

Growing cowpea — water

Water needs

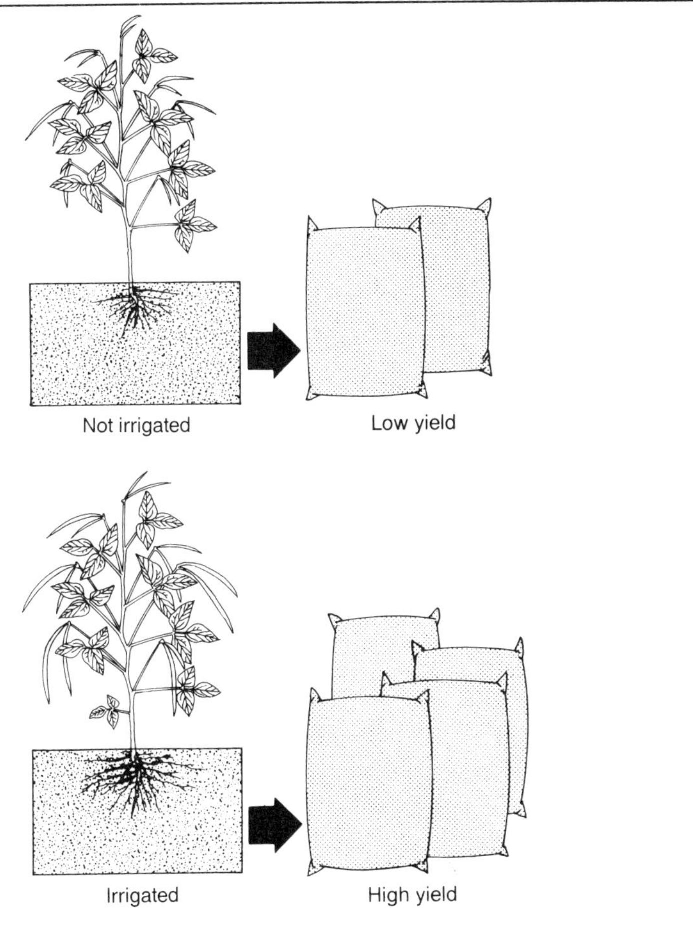

- Although cowpea can stand drought, water supplied during critical stages will give higher yields.

When water is most needed

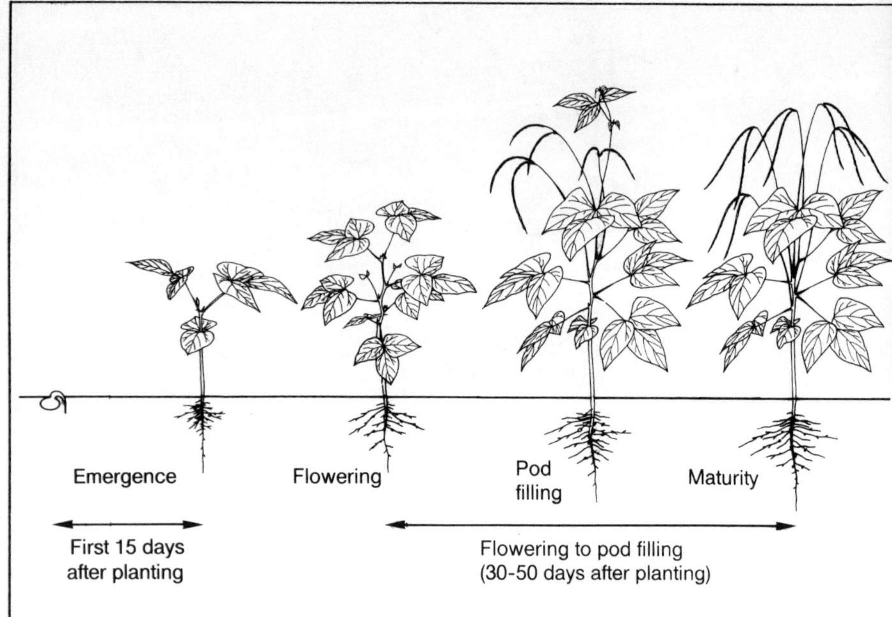

- Cowpea needs water most
  - at planting
  - during early seedling growth
  - during flowering
  - during pod filling.

Effects of drought

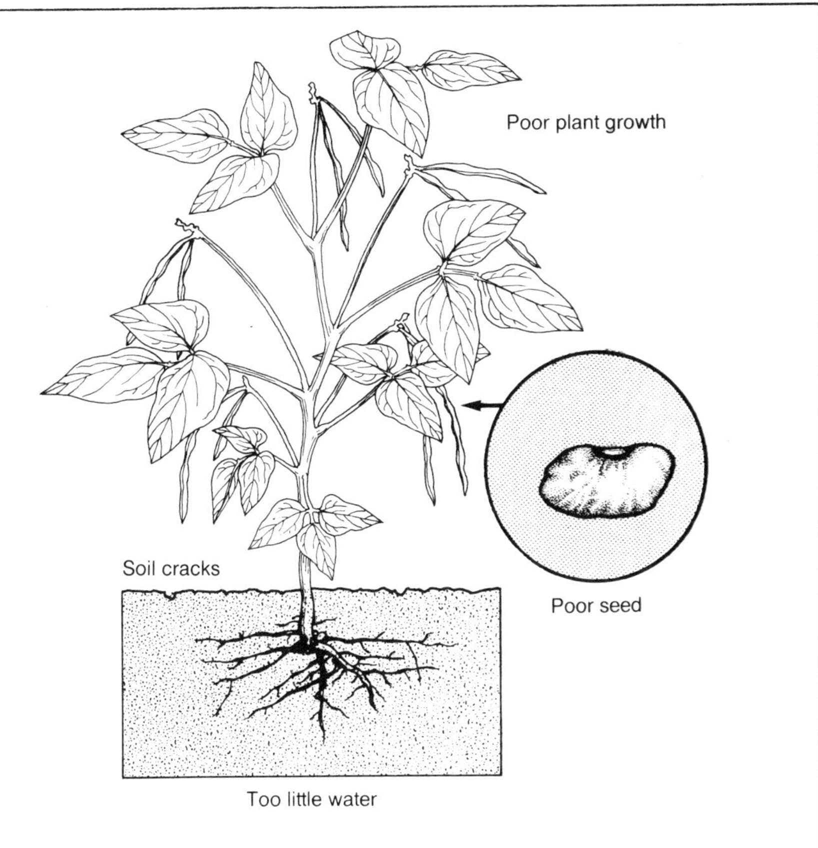

- Drought can
  - stunt plant growth
  - reduce nodulation and nitrogen fixation
  - decrease protein content of seed
  - reduce total yield.

Effects of too much water

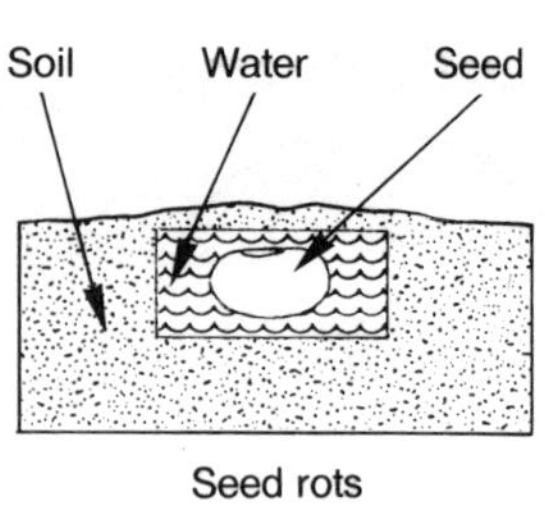

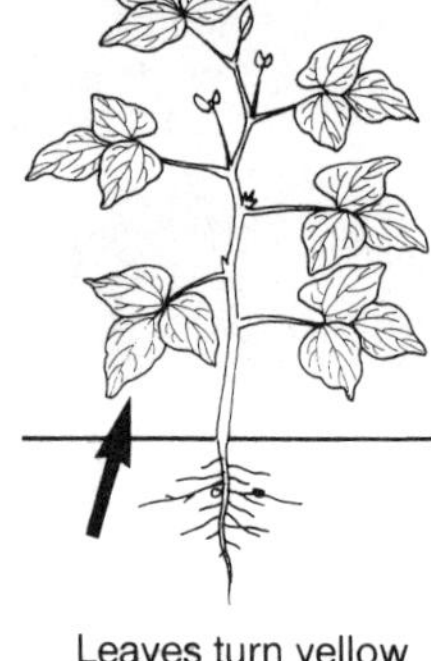

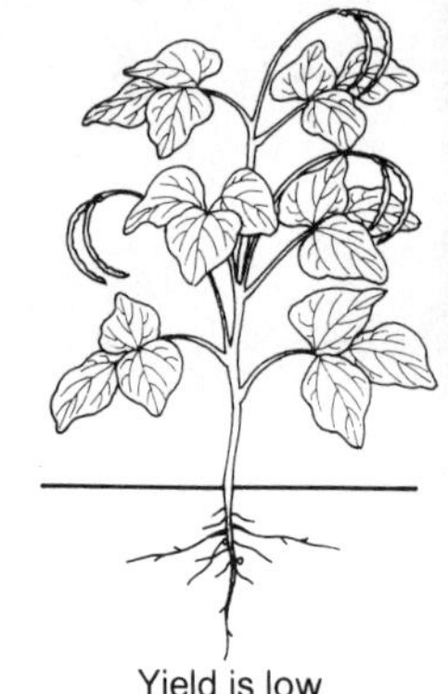

- Too much water can
  - delay germination and rot the seed
  - reduce nitrogen fixation
  - reduce total yield.

Growing cowpea — choosing the right variety

Choosing the right variety — before rice

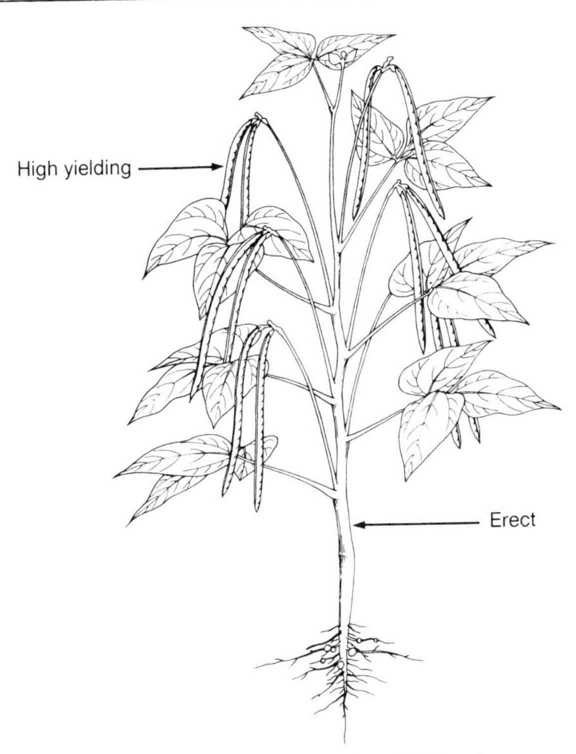

Early varieties planted before rice

- Cowpea varieties planted before rice should be
  — erect growing, with most pods maturing at the same time
  — early-maturing
  — able to stand drought during early growth stages
  — able to stand excess water during flowering and pod filling.

Choosing the right variety — after rice

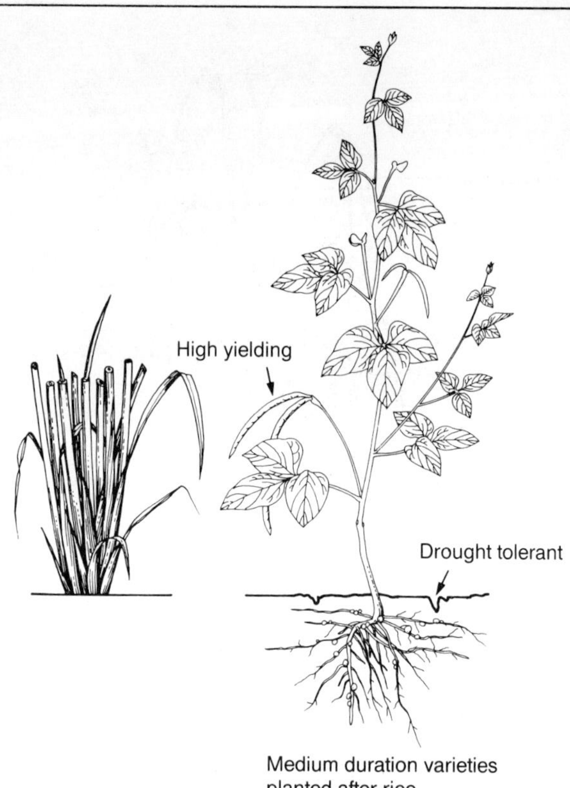

- Cowpea varieties planted after rice should be
  - indeterminate types, with pods maturing over several days
  - medium-duration
  - resistant to wilt disease
  - able to stand excess water during early growth
  - able to stand drought at flowering and pod filling.

Choosing the right variety after rice

- Relay-cropped varieties should have especially vigorous seedlings.

Choosing the right variety — pest and disease resistance

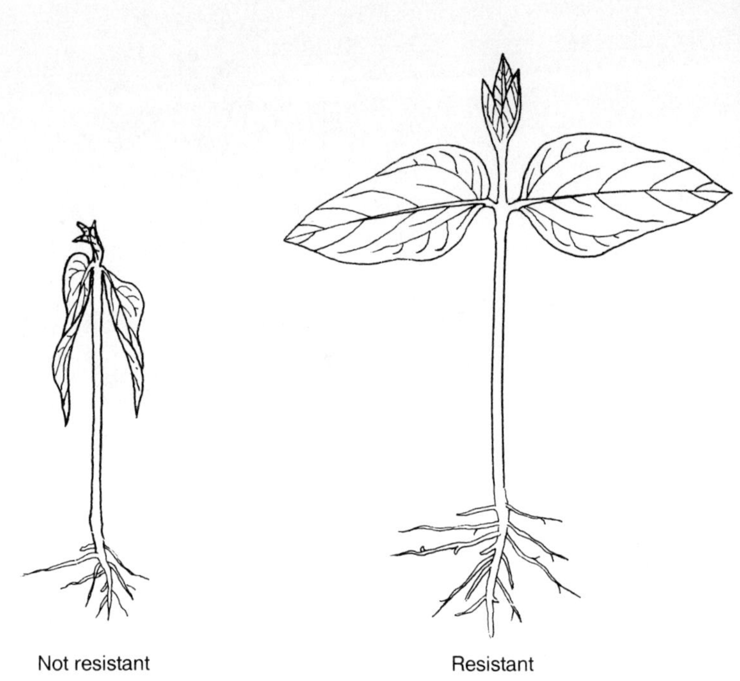

- Some cowpea varieties resist insect and disease damage better than others.
- Choose varieties that are least damaged by the major pests and diseases in your area.

Growing cowpea — tillage and planting

Preparing the land — high tillage

- High tillage is common in irrigated areas where water is easily available.
- High tillage
  - airs the soil
  - help seeds germinate and roots grow deep
  - controls weeds.
- But high tillage
  - is costly
  - delays planting
  - dries out the soil.

Preparing the land — zero tillage

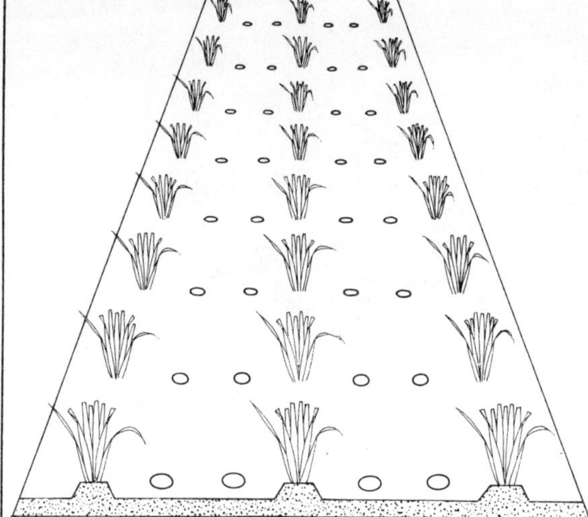

Zero tillage

No plowing

No harrowing

- Zero tillage is common in rainfed areas, especially after lowland rice.
- Zero tillage
  - saves labor and costs
  - allows planting at once
  - makes full use of soil moisture.
- But zero tillage
  - does not air the soil
  - lets weeds grow
  - does not help roots grow deep.

Planting system

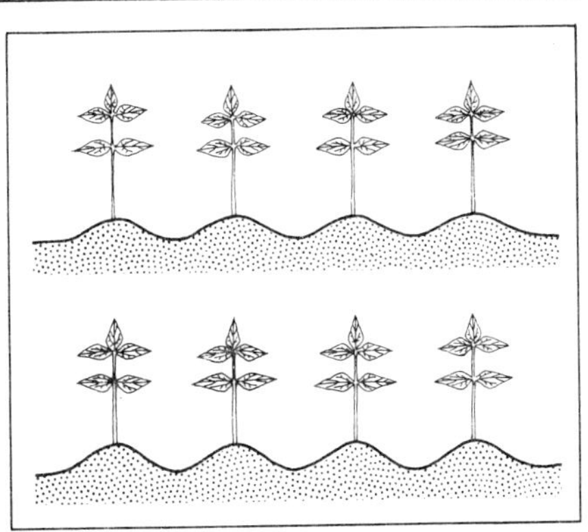

Sole cropping
before or after rice

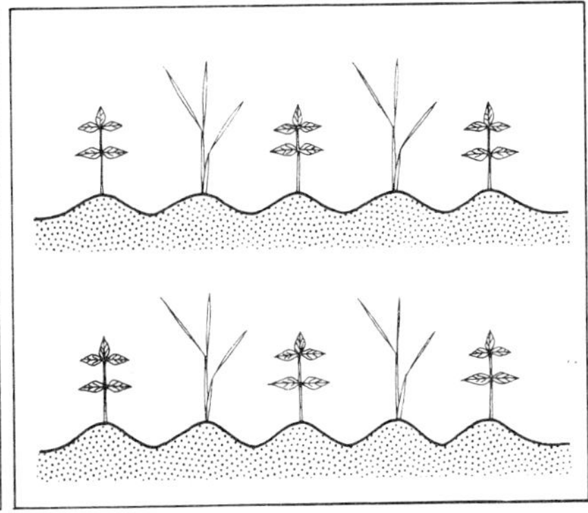

Intercrop with
upland rice

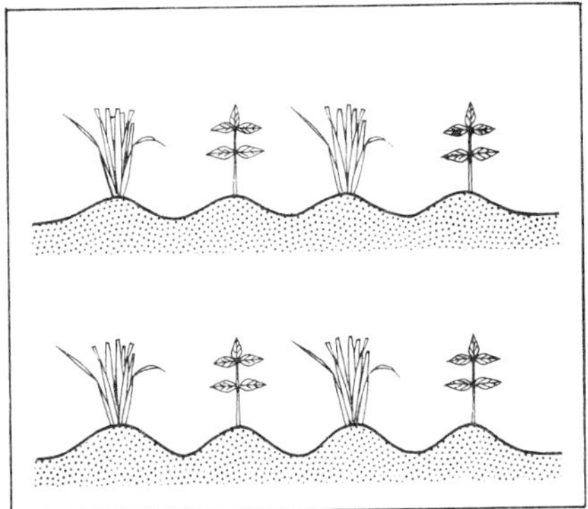

Relay crop
in standing water

- Cowpea can be planted as a sole crop before or after rice.
- It can also be relay cropped or intercropped with upland rice.

When to plant as a sole crop

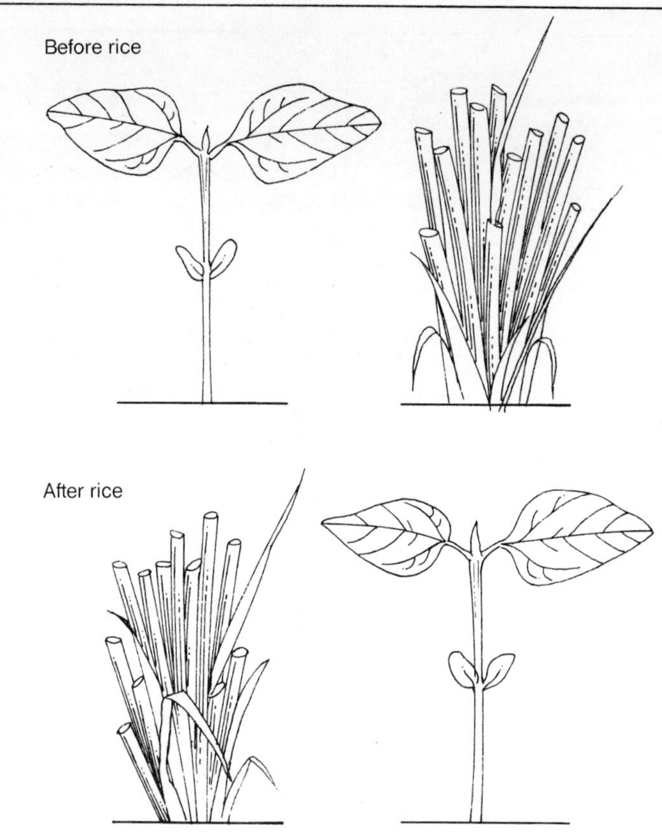

- As a sole crop before rice, plant cowpea at the start of the rainy season, in May.
- After rice, plant cowpea in November, after the harvest of rice.

When to plant as a relay crop

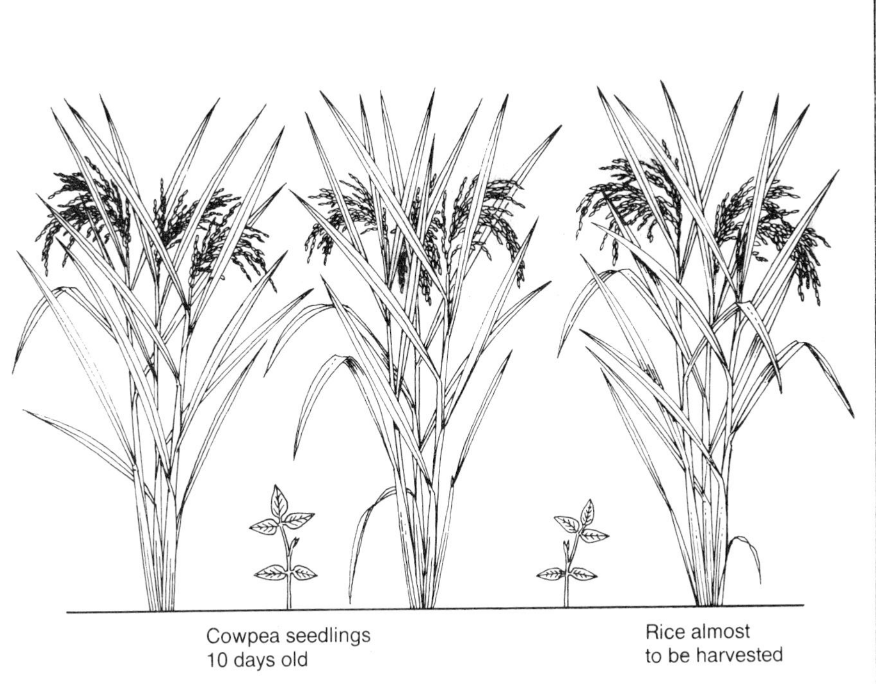

- For a relay crop, plant cowpea in standing rice, about 10 days before the rice harvest.

Row spacing — sole crop

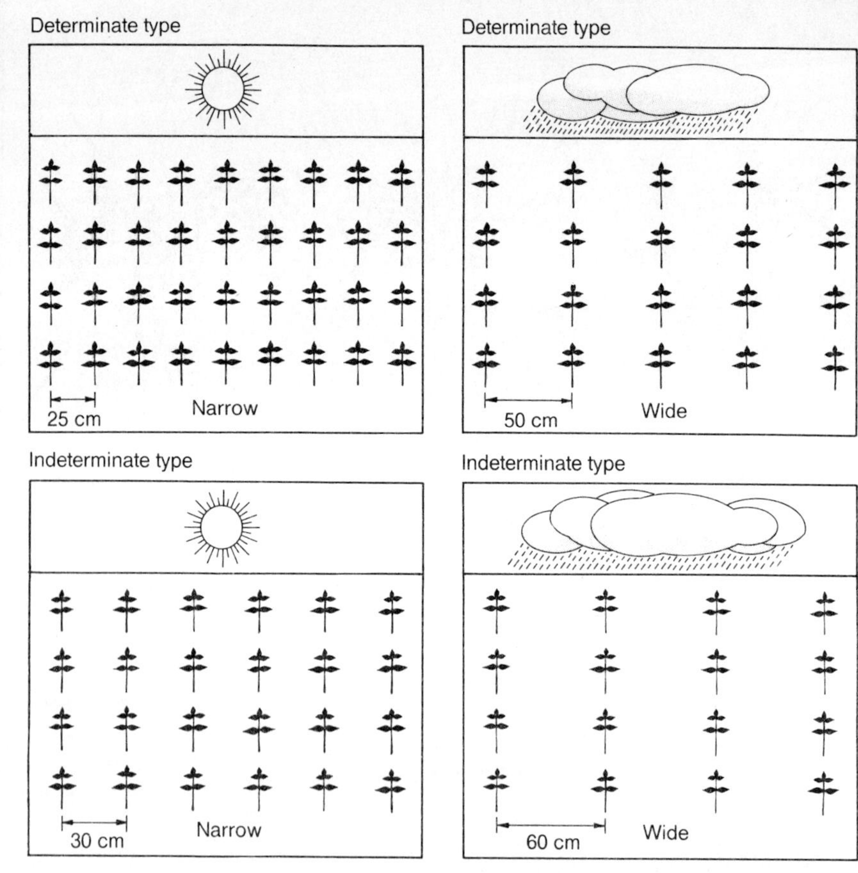

- Space between rows varies with plant type and season.
- Use narrow row spacing for determinate types and in the dry season.
- Use wide row spacing for indeterminate types and in the wet season.

Row spacing — intercrop

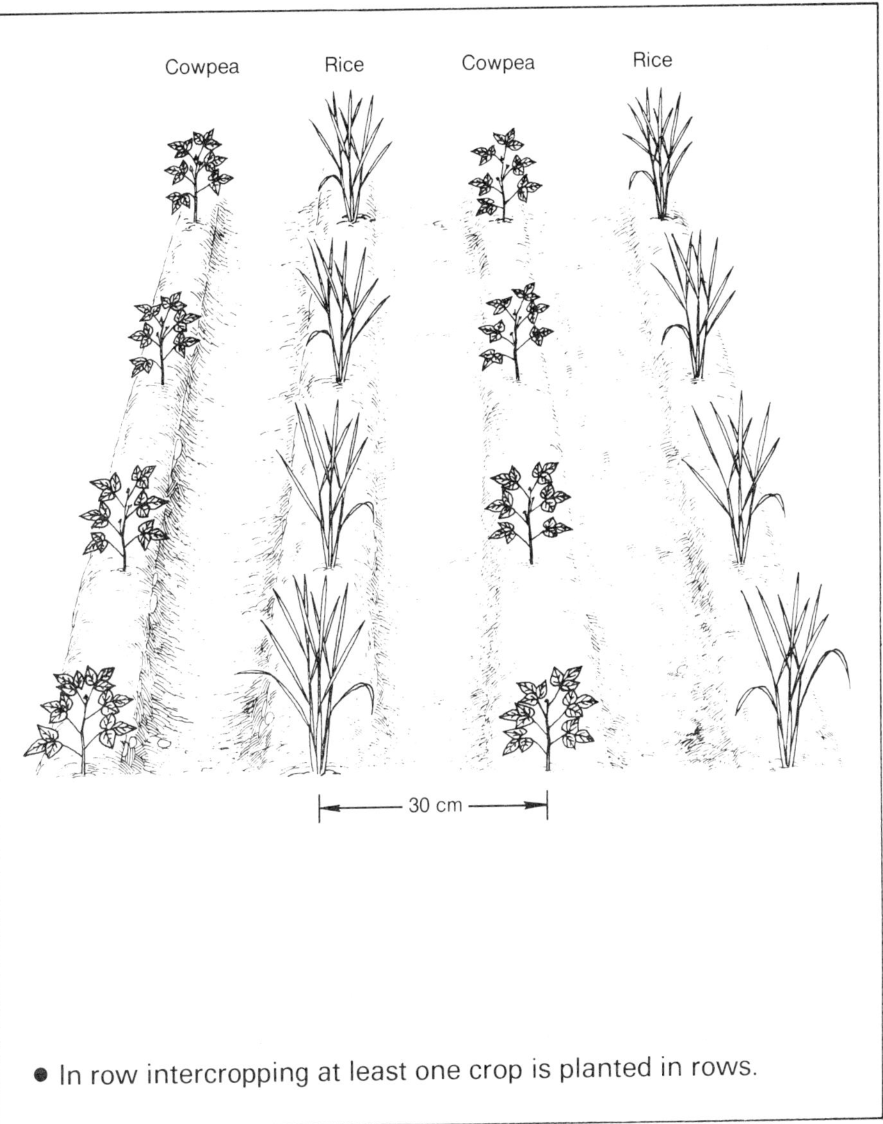

- In row intercropping at least one crop is planted in rows.

Mixed intercropping

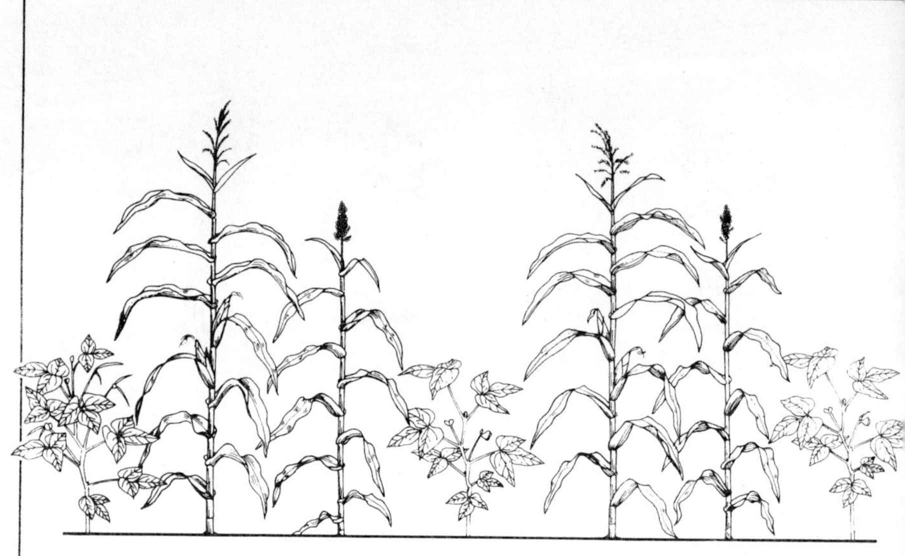

- Mixed intercropping uses no distinct row arrangement or row spacing.
- As a fodder crop, cowpea is often mixed intercropped with cereal crops.

Planting method

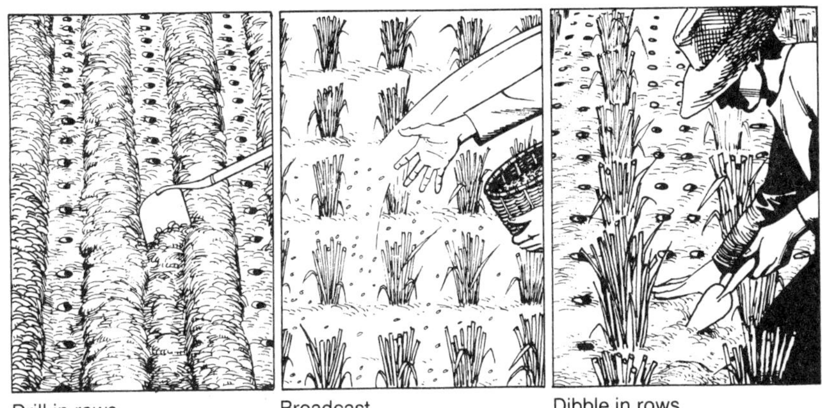

- Drill seed in rows by hand or by animal-drawn seeder.
- Dibble seed at the base of rice stubble after rice harvest.
- For mixed or relay crops, broadcast seed in tilled fields and cover with soil. Or broadcast without tilling, directly in wet fields.

Planting depth

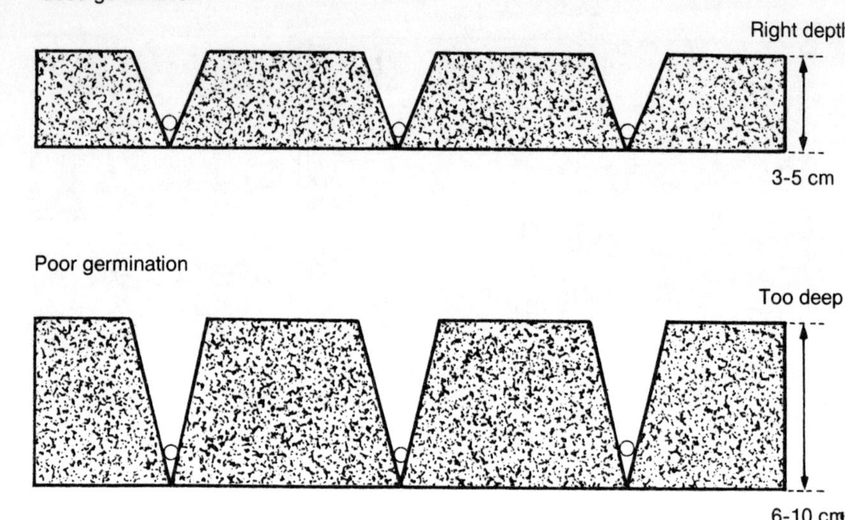

- Sowing 3 to 5 cm deep is good for most varieties.
- Planting more than 6 cm deep delays emergence. Seed may rot and plant stands will be uneven.

Seeding rate

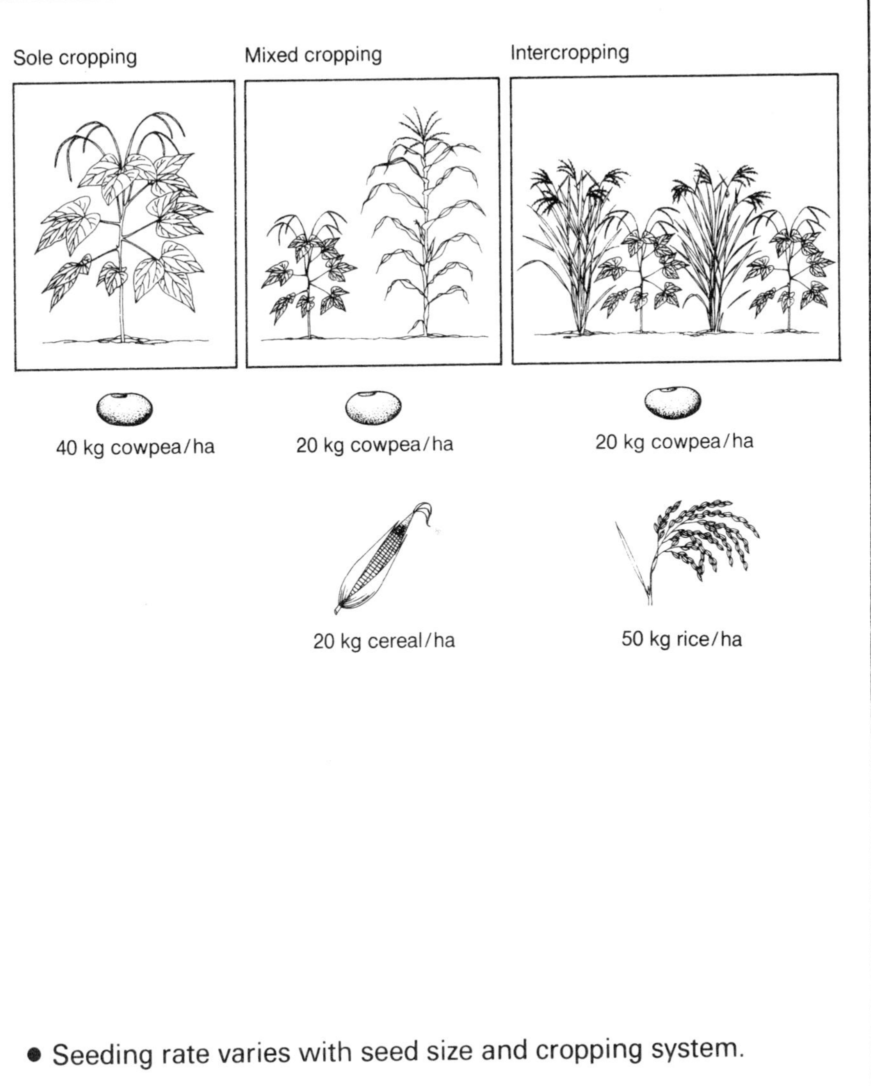

- Seeding rate varies with seed size and cropping system.

Plant density

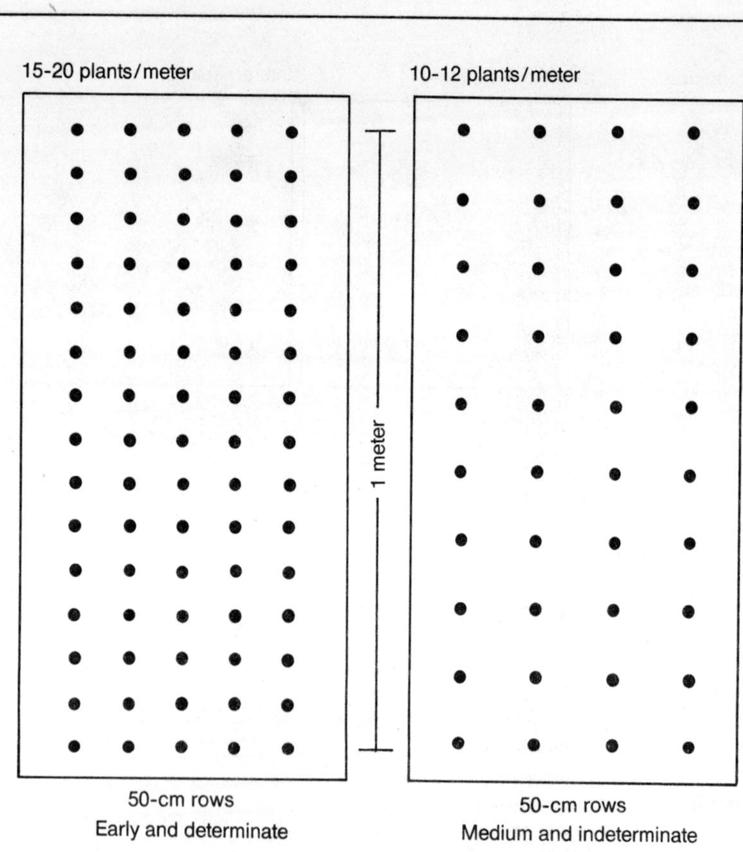

- For a sole crop of cowpea the best plant density is
  - 15 to 20 plants per meter row for early, determinate varieties.
  - 10 to 12 plants per meter row for medium-duration, indeterminate varieties.

Fertilizer and lime

Fertilizer needs

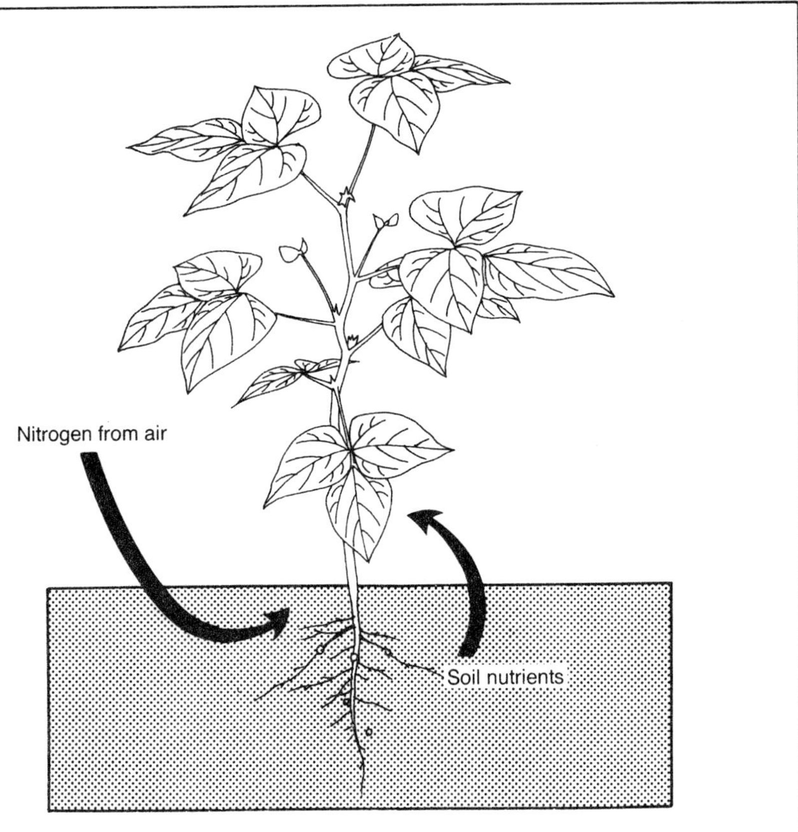

- The cowpea crop usually does not need fertilizer. It uses nitrogen from the air and other nutrients left in the soil from the previous crop.
- In poor soils, however, adding fertilizer will improve yields.

Organic fertilizer

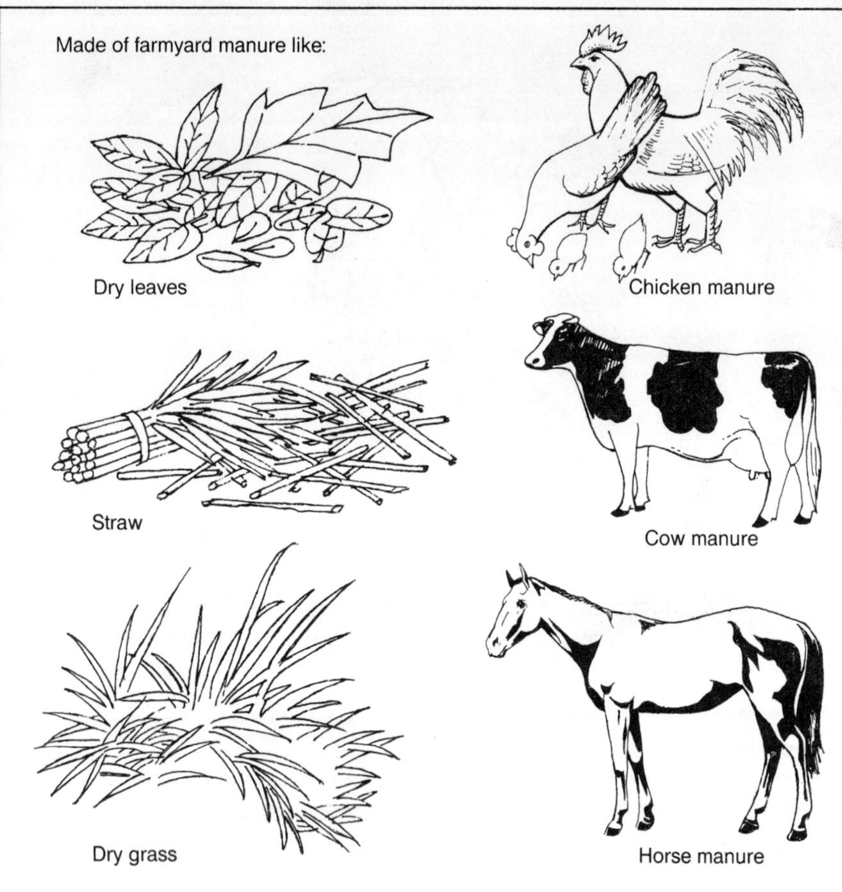

- Add organic fertilizer in any amount possible.
- Large amounts are needed to improve yields significantly.
- But even smaller amounts improve soil structure and help plant growth.

Fertilizer — nitrogen

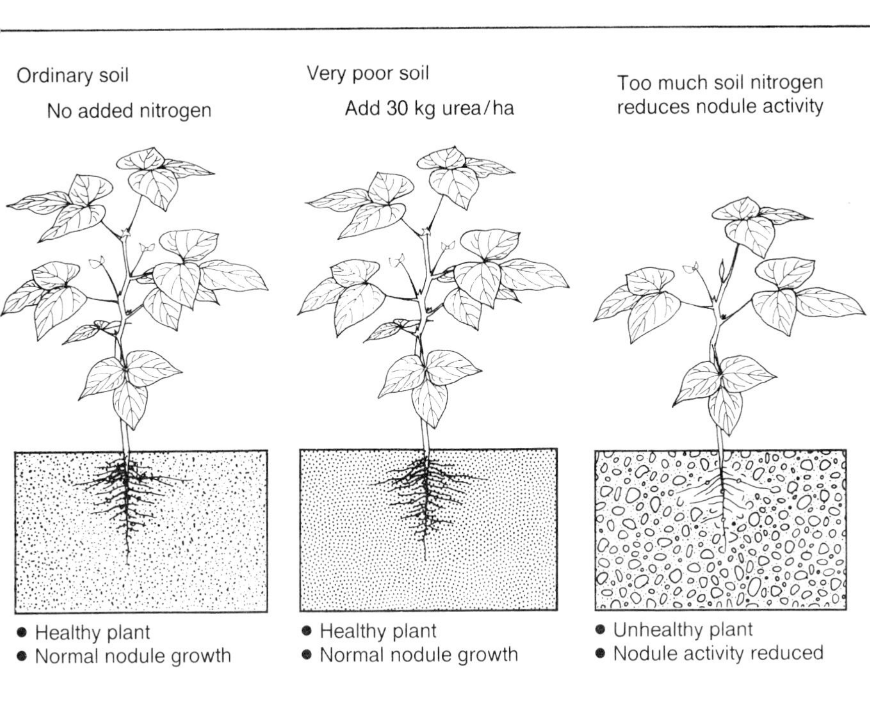

- Cowpea needs no added nitrogen fertilizer.
- In very poor soils, add 30 kg urea per hectare at planting to help start the crop.

Fertilizer — phosphorus

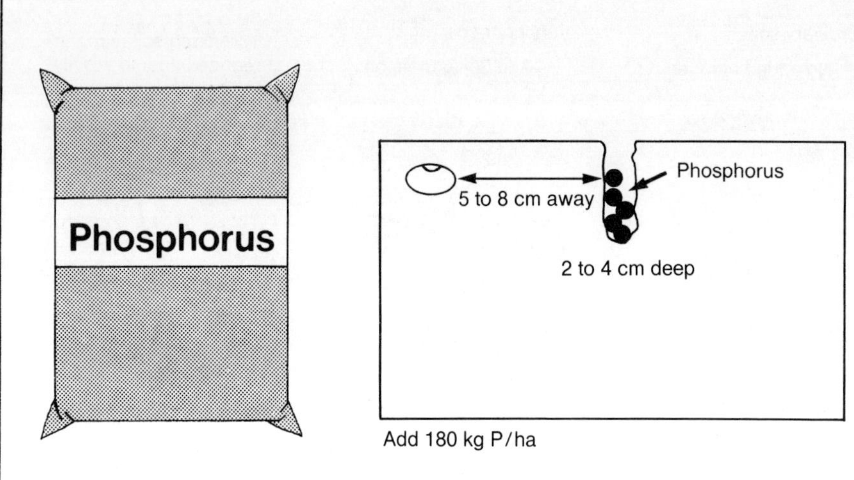

Add 180 kg P/ha

- Phosphorus is needed for good nodulation and nitrogen fixation.
- If soil is low in phosphorus, add 180 kg single super-phosphate at planting time.

Fertilizer — potassium

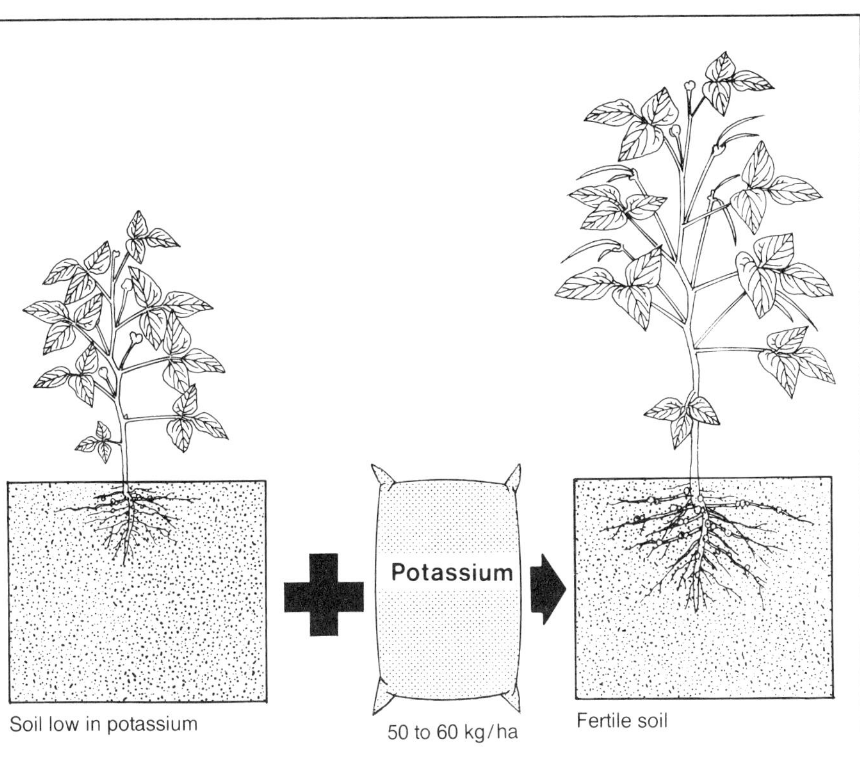

- Most tropical soils have enough potassium and rarely need added potash.
- If soil tests low in potassium, add 50 to 60 kg potash per hectare.

Fertilizer — micronutrients

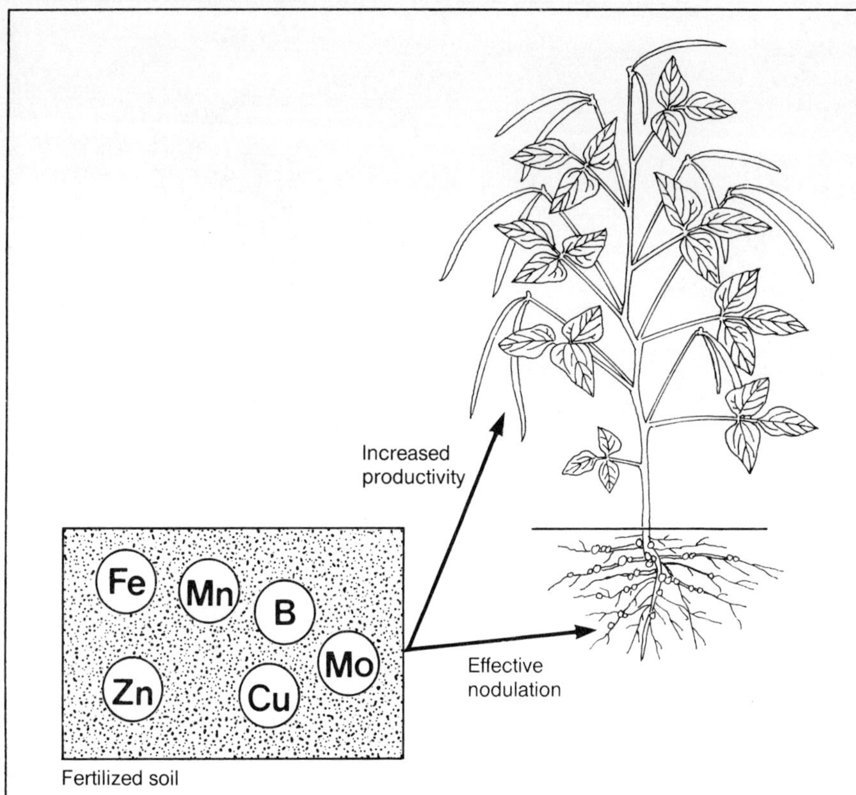

- Cowpea also needs micronutrients for proper growth and good yields.

Lime

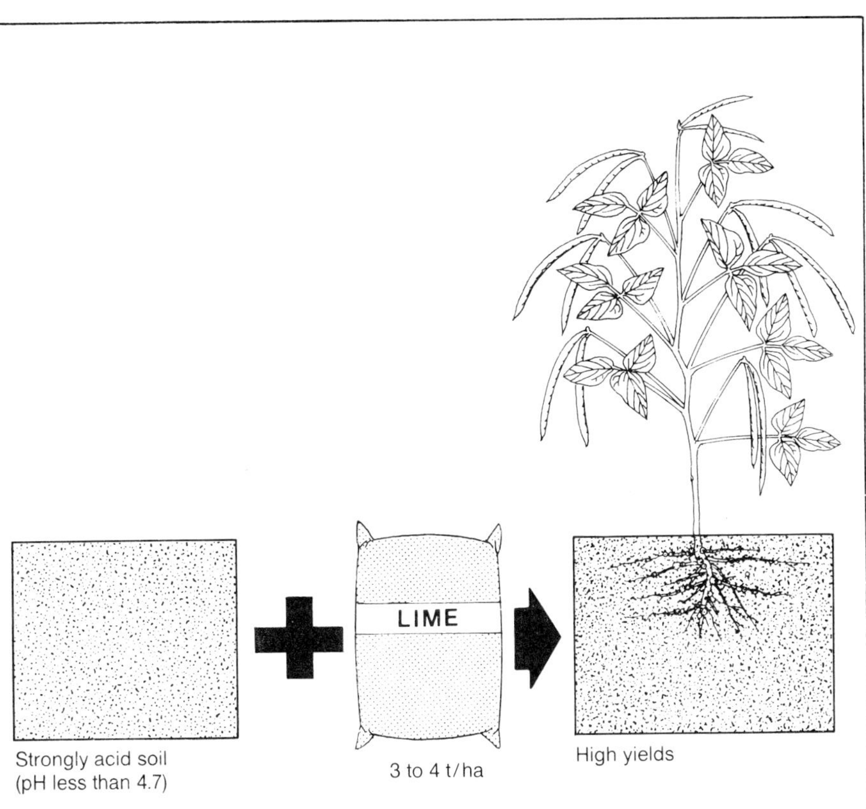

- Cowpea can usually stand acid soils. But strongly acid soils, with pH less than 4.5, need added lime to give high yields.

Harvesting and storage

When to harvest — vegetable

- For use as a green vegetable, hand-pick cowpea pods within 12 to 14 days after flowering, when pods are still tender.
- Pick every 3 or 4 days after that.

When to harvest — seed

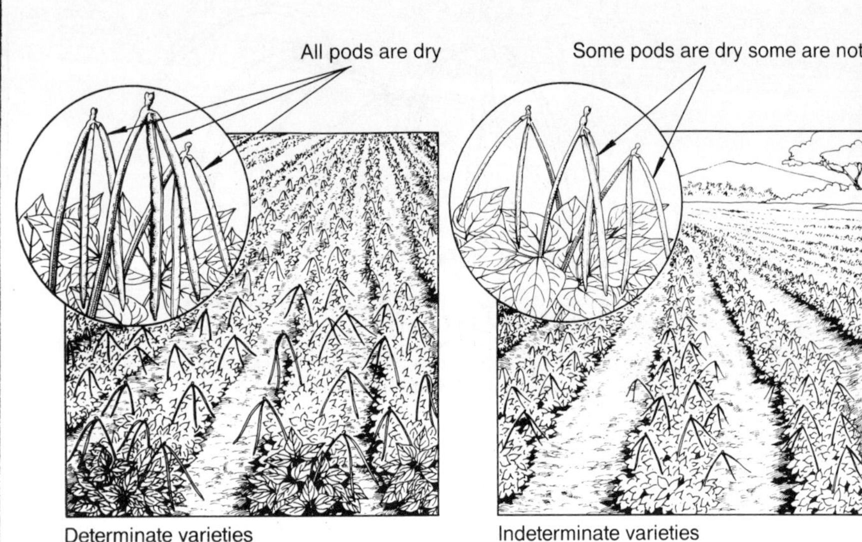

Determinate varieties
Harvest when 85-90% pods are dry

Indeterminate varieties
Harvest dry pods only. Two or
three pickings needed.

- Varieties maturing evenly can be harvested within 20 to 25 days after flowering, when most of the pods are dry.
- For varieties maturing unevenly, two or three pickings are needed.

When to harvest — fodder

- Harvest cowpea grown for fodder at flowering to early pod formation stage for maximum dry matter and crude protein.
- Cut at the base of the plant.

Seed drying

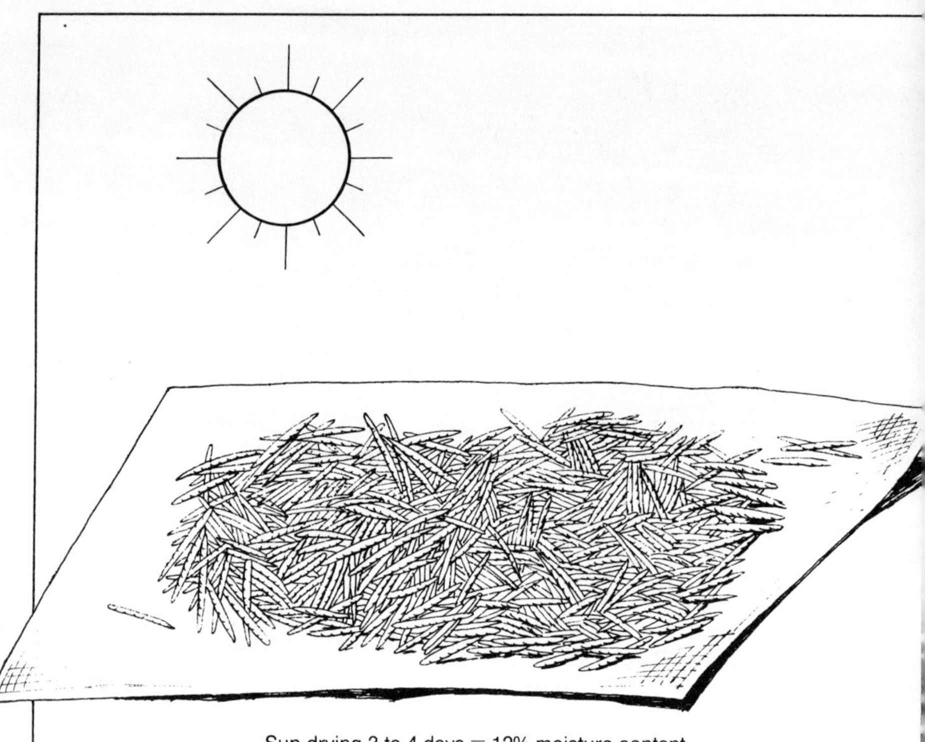

Sun drying 3 to 4 days = 12% moisture content

- Harvested pods are dried 3 to 4 days under the sun or in a dryer until the moisture content is about 12 percent.

Threshing

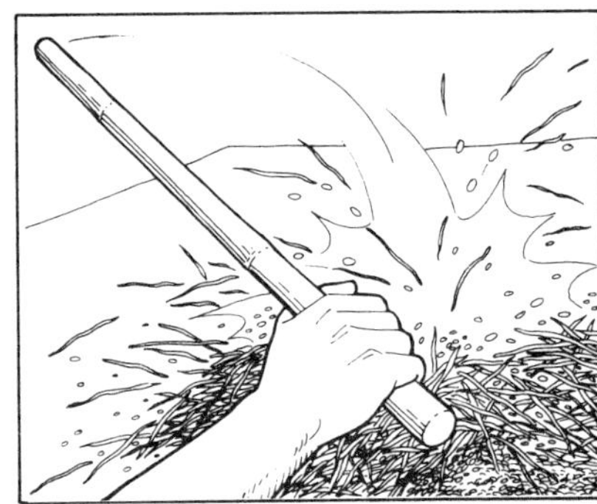

Hand threshing

Animal threshing

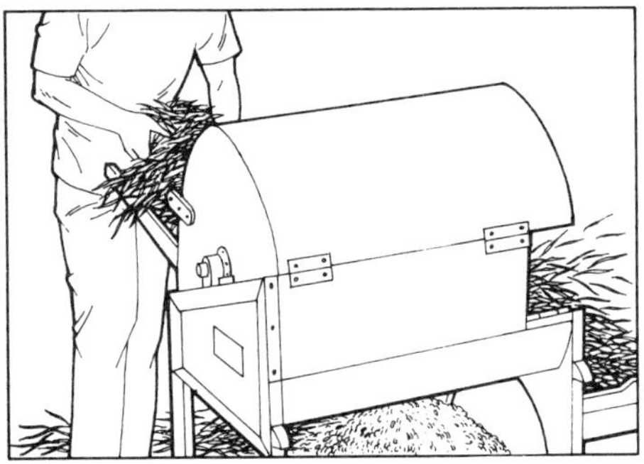

Machine threshing

- Hand-threshing is done by beating with a stick.
- Sometimes cattle may be used to trample dry pods.
- For large-scale production, cowpeas can be machine-threshed.

Storage

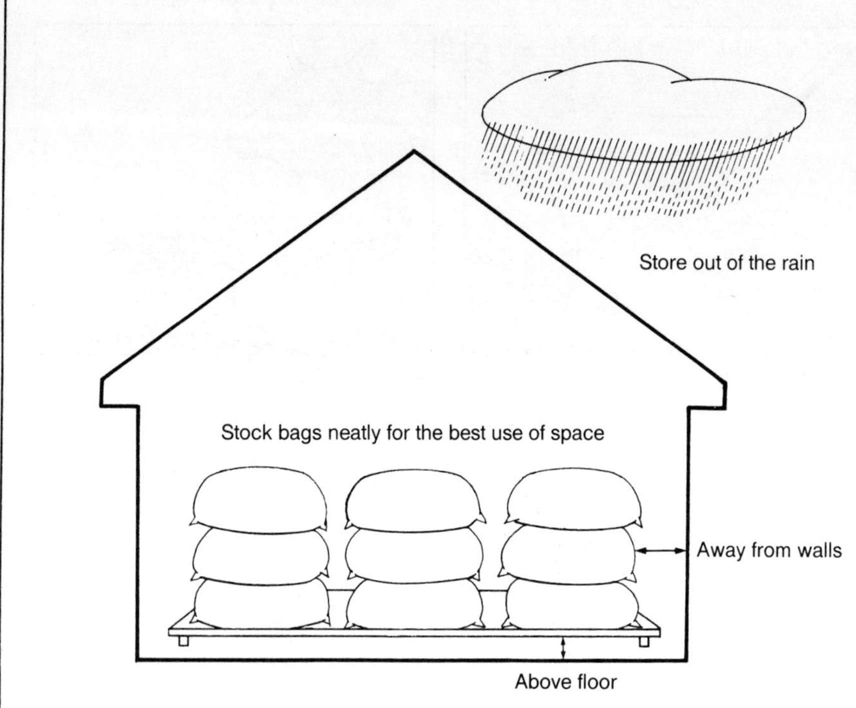

- Seed for storage should be sundried or machine-dried thoroughly.
- For cold storage, set temperature at 6 to 8°C.

Controlling storage pests

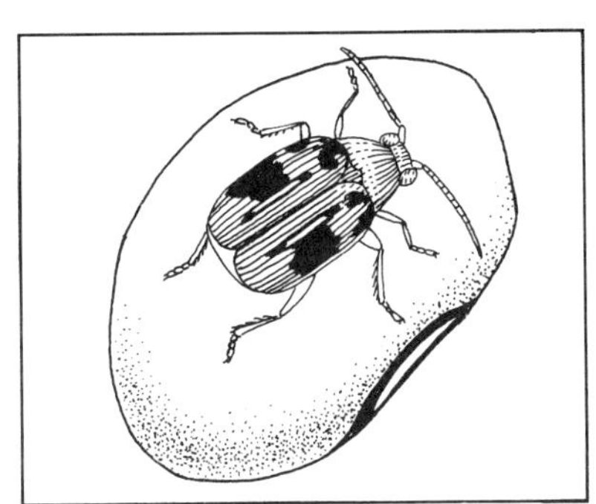

Bean weevil

Weevil damage

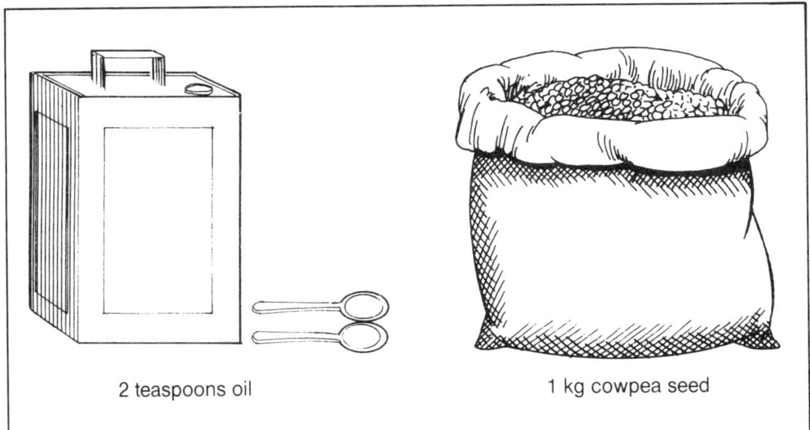

2 teaspoons oil

1 kg cowpea seed

For protection against bean weevil

- The bean weevil can severely damage stored cowpea seed.
- Mix seeds with vegetable oil to protect against this pest.

Increasing yields and profits

Increasing yields and profits — yield components

Yield components

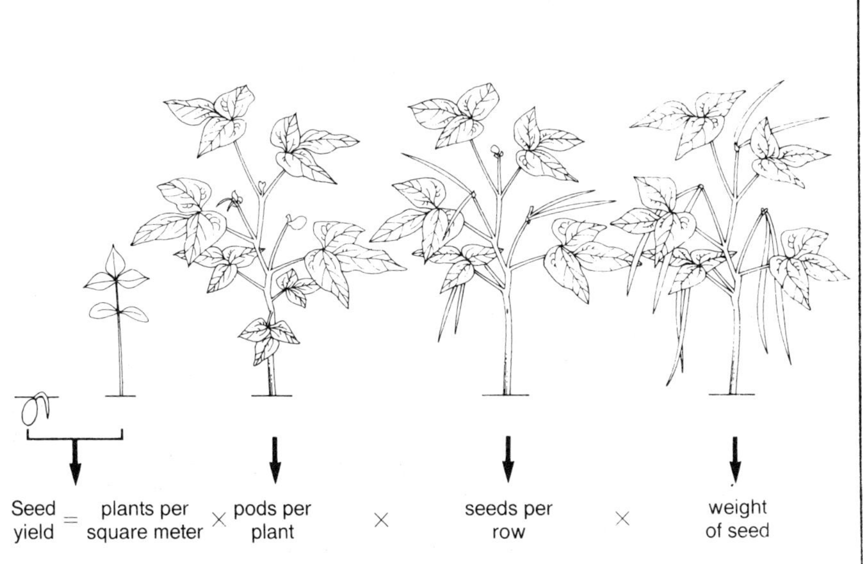

- Each yield component contributes to the total yield. Reducing any component reduces yields.
- Good management at all stages is needed for high yields, because growing conditions affect each stage of development.
- Some yield components are determined more by variety than by environment.

Yield components — plants per unit area

Plants per square meter determine pods per unit area

- Number of plants bearing mature pods will determine total number of pods.

Yield components — pods per plant

Number of pods per plant is determined from:

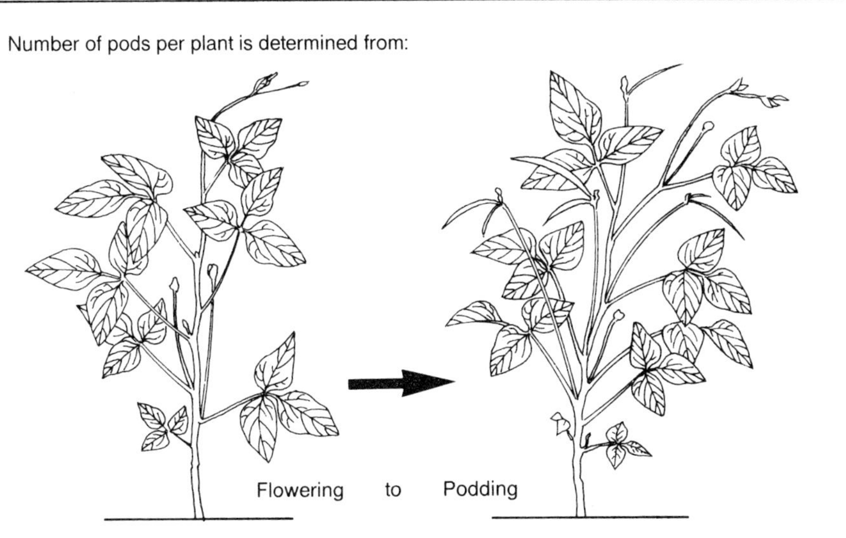

Number of pods that mature depends on:

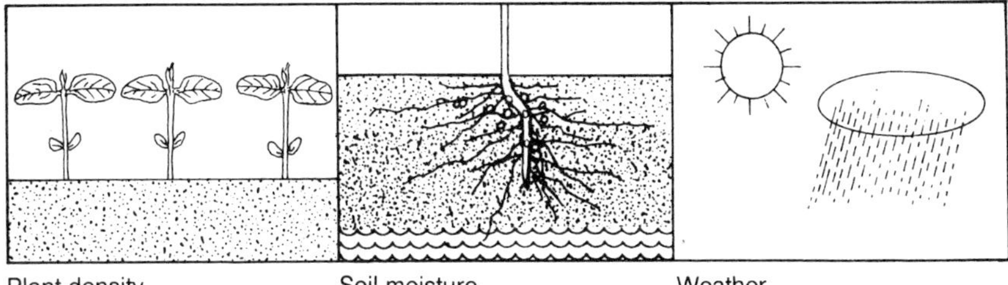

- The number of pods per plant is the most important yield component.
- It is the most affected by growing conditions: plant density, soil moisture, and weather.

Yield components — seeds per pod

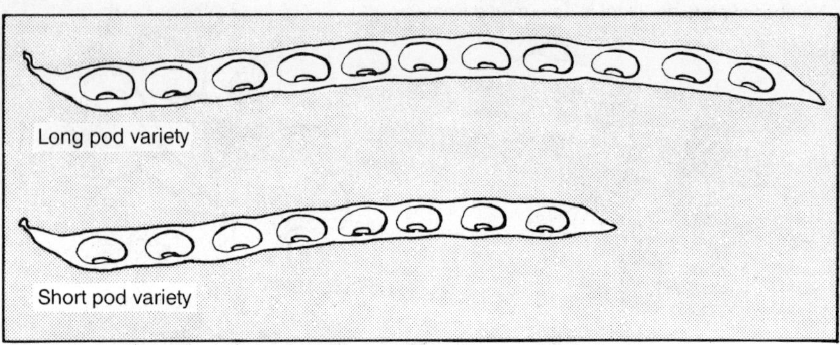

Adequate water and nutrient supply

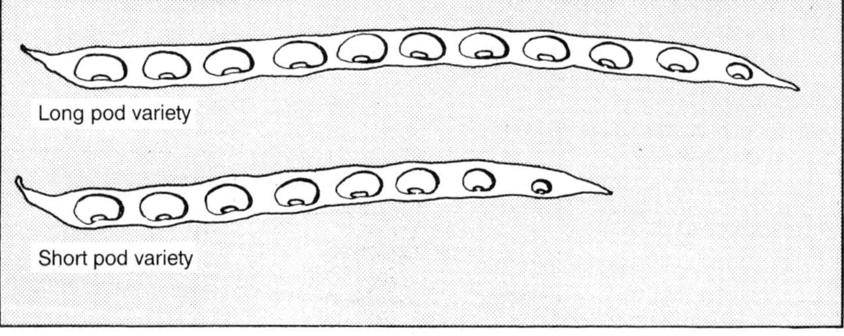

Poor water and nutrient supply

- The number of seeds per pod is determined at flowering, when male pollen cells are transferred to the ovules in the pod.
- Fertilized ovules will develop into seeds.

Yield components — seed weight

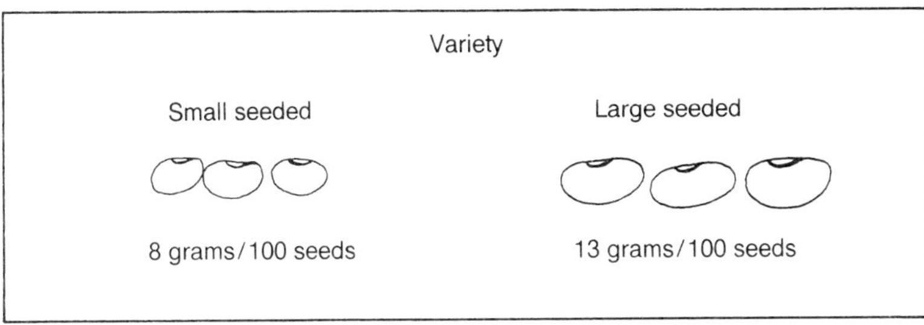

Good soil moisture and nutrient supply ensure
proper seed filling

Poor soil moisture and nutrient supply

- Seed weight is determined during pod filling.
- It depends on variety, soil moisture, and nutrient supply.

Increasing yields and profits — production factors

Production factors

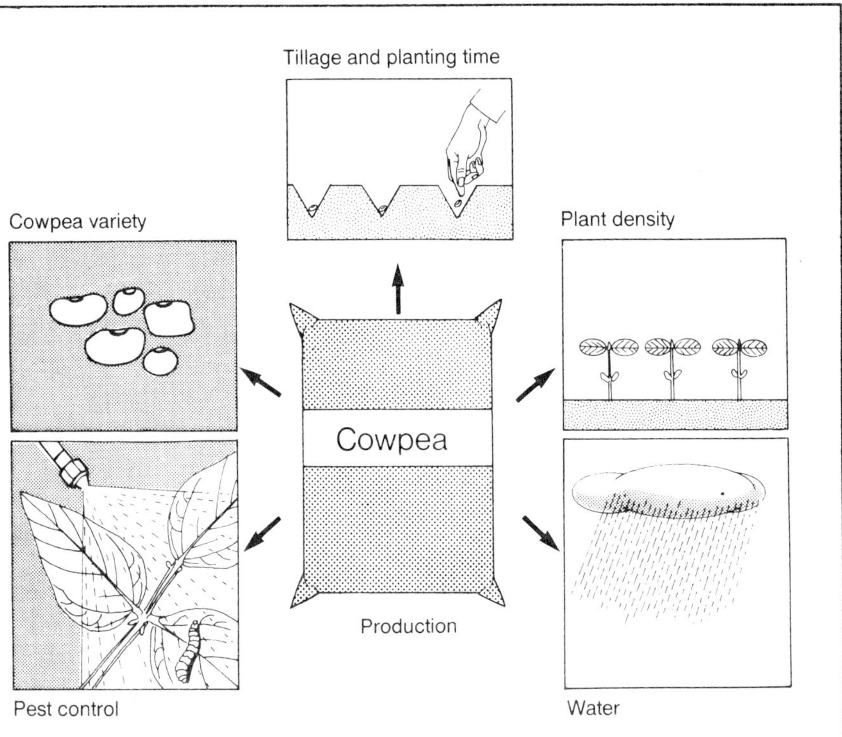

- Cowpea is a low-cost crop to grow. With the right combination of production factors, yields and profits can be high.
- The right combination varies with season, location, and growing conditions.

Making the most of soil moisture — tillage and planting time

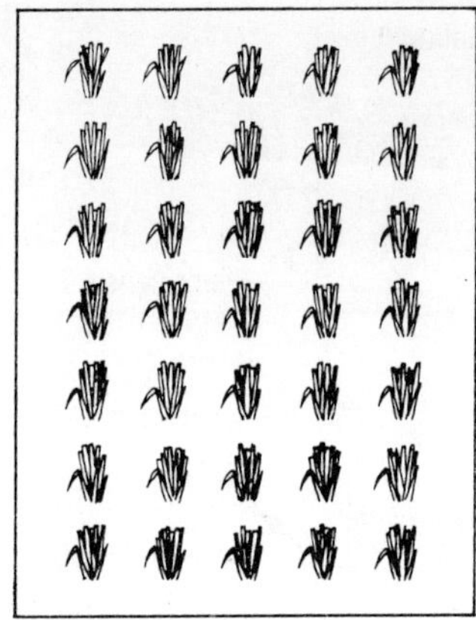

Use zero tillage

Use narrow rows

|←→| 25 cm

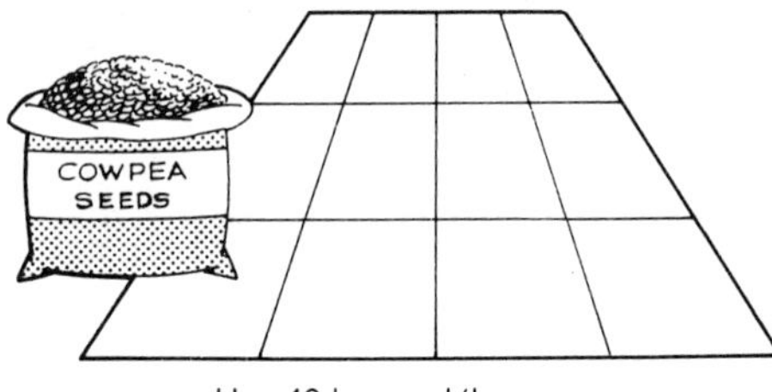

Use 40 kg seed/ha

- In rainfed crops, making the best use of soil moisture is the key to high yields.
- Plant the cowpea crop at once after the rice harvest. Or plant as a relay crop in standing rice 10 days before harvest.
- Use zero tillage and narrow spacing between rows. High tillage and wide row spacing dry out the soil.

Making the most of soil moisture — variety

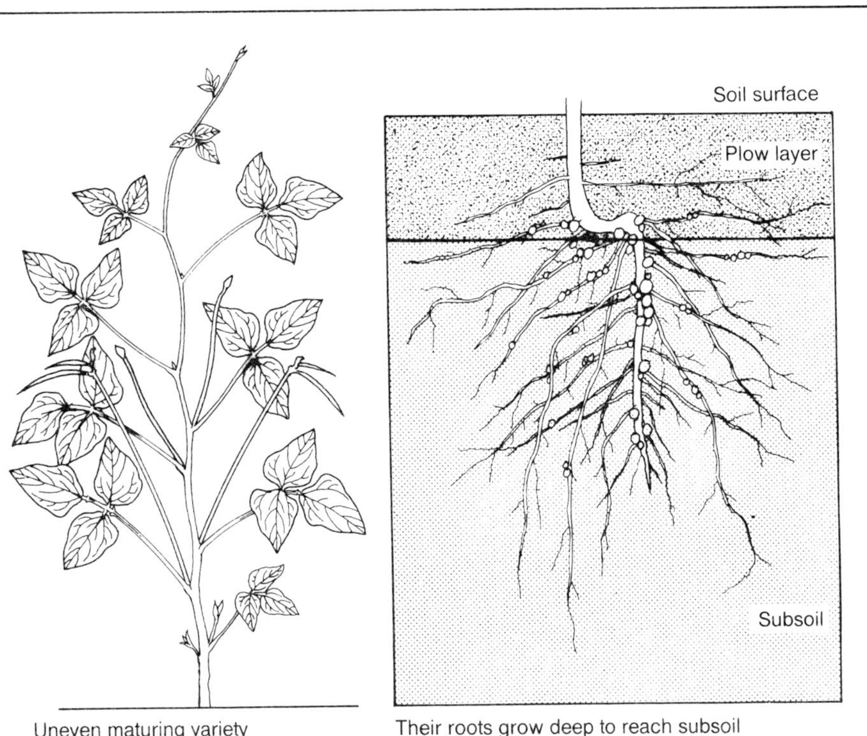

Uneven maturing variety

Their roots grow deep to reach subsoil

- Plant indeterminate varieties that mature unevenly.
- They yield more than determinate varieties in the dry season.
- At all times, plant varieties resistant to insect pests and diseases.

Making the most of soil moisture — fertilizing and weeding

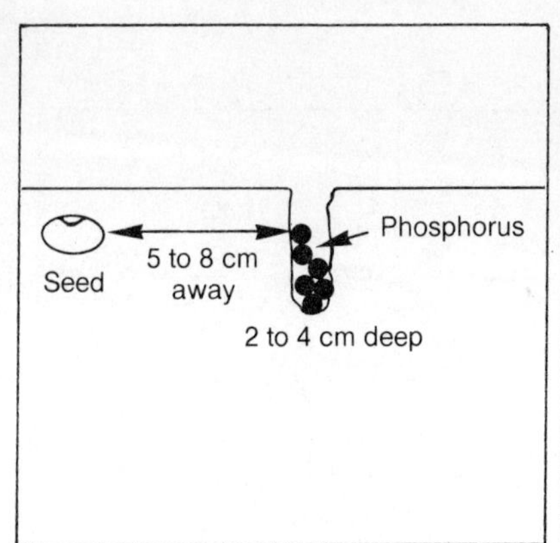

Add 180 kg of single superphosphate at planting time

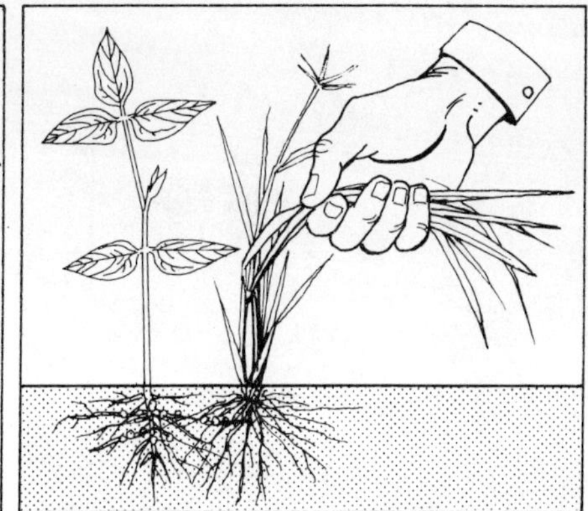

Remove weeds that steal crop nutrients

- Add phosphorus at planting time for good nodule growth and nitrogen fixing.
- Weed at least twice during the first 40 days.

Increasing yields — using irrigation

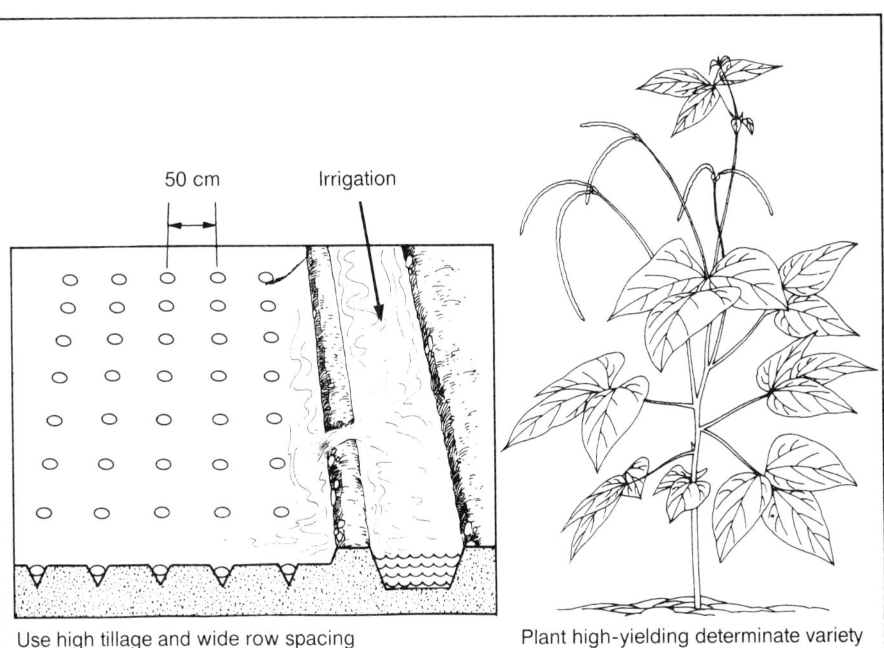

Use high tillage and wide row spacing

Plant high-yielding determinate variety

- Where water is available, use high tillage and wide row spacing. Irrigate during early growth stages and flowering and pod filling.
- Grow determinate, high-yielding varieties that mature evenly.

Yield reducers — weeds

Yield loss to weeds

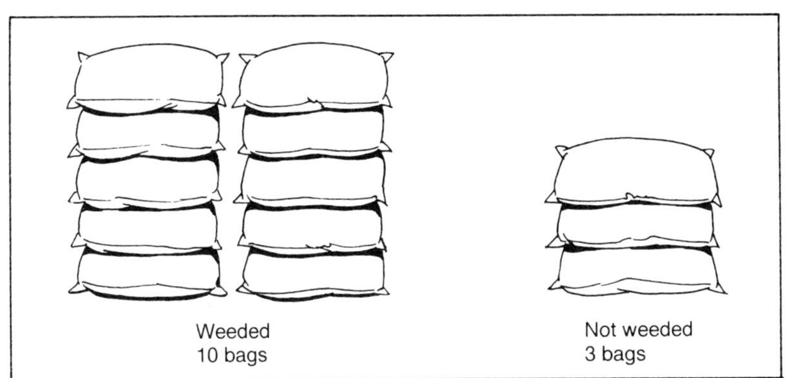

Seed yield

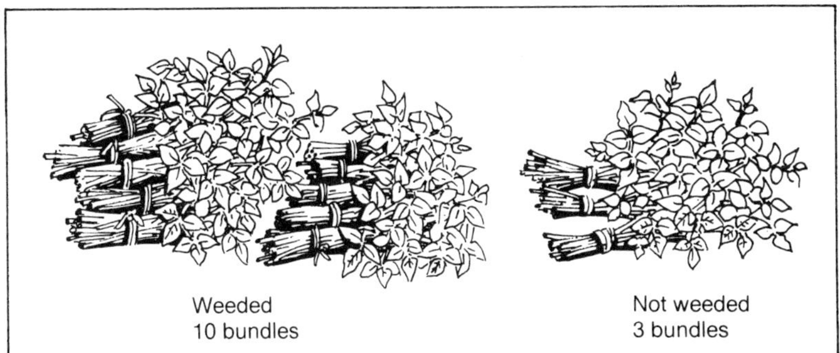

Fodder yield

- Uncontrolled weeds may reduce cowpea yields by 60 to 70 percent. Seed yields may come down from 1000 to 300 kg per hectare. Fodder yields may come down from 10 tons to 3 tons per hectare.

Weeds compete with cowpea

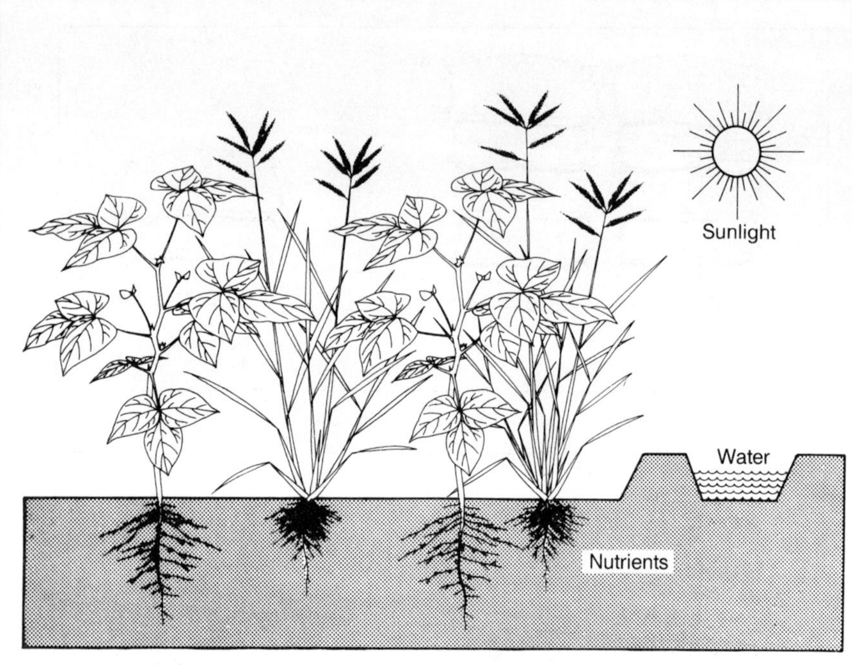

- Weeds compete with cowpea for soil nutrients, soil water, and sunlight.

Weeds affect seedling growth

Weeds harm most from emergence to 40 days later

- Weeds do most harm in the first 40 days after planting.
- After the crop has flowered, weeds are not as damaging as at early stages.

Controlling weeds — by handweeding

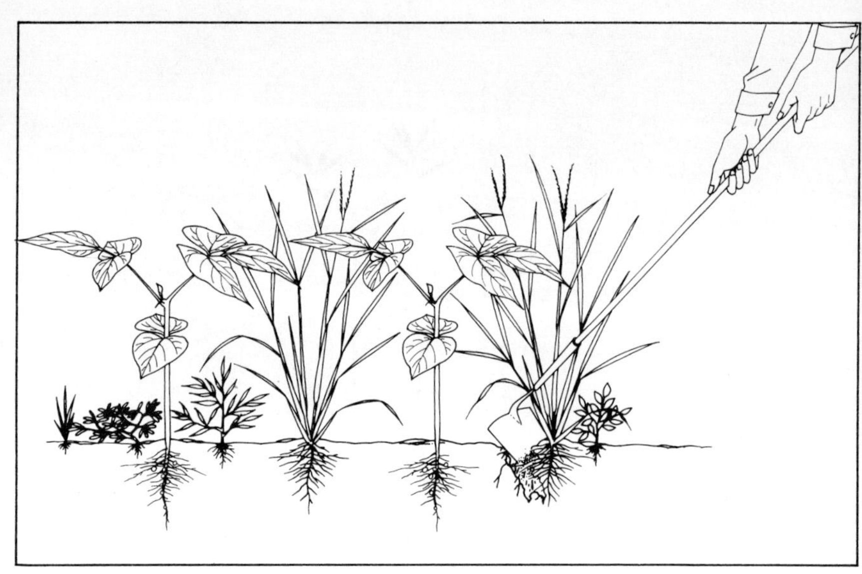

- Weeding with a hand hoe is the most common practice among farmers.
- For best yields, weed 2 weeks after planting and just before flowering.

Controlling weeds — using cultural practices

Tractor

Animal-drawn desi plow

Hoe

- Two or three intercultivations with a hoe or animal-drawn tool or a tractor will control cowpea weeds.
- Close plant spacing keeps down weeds.

Controlling weeds — using herbicides

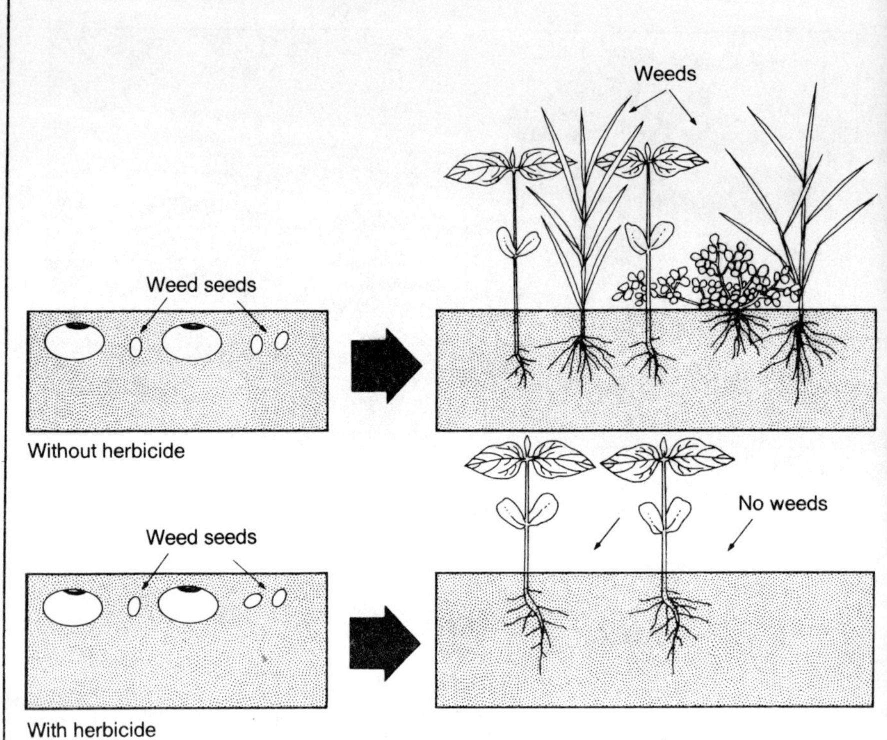

- For large-scale cowpea production, chemicals can be used to control weeds.
- If soil is moist, use herbicide before weeds emerge, just after planting cowpea.

Common cowpea weeds — grasses

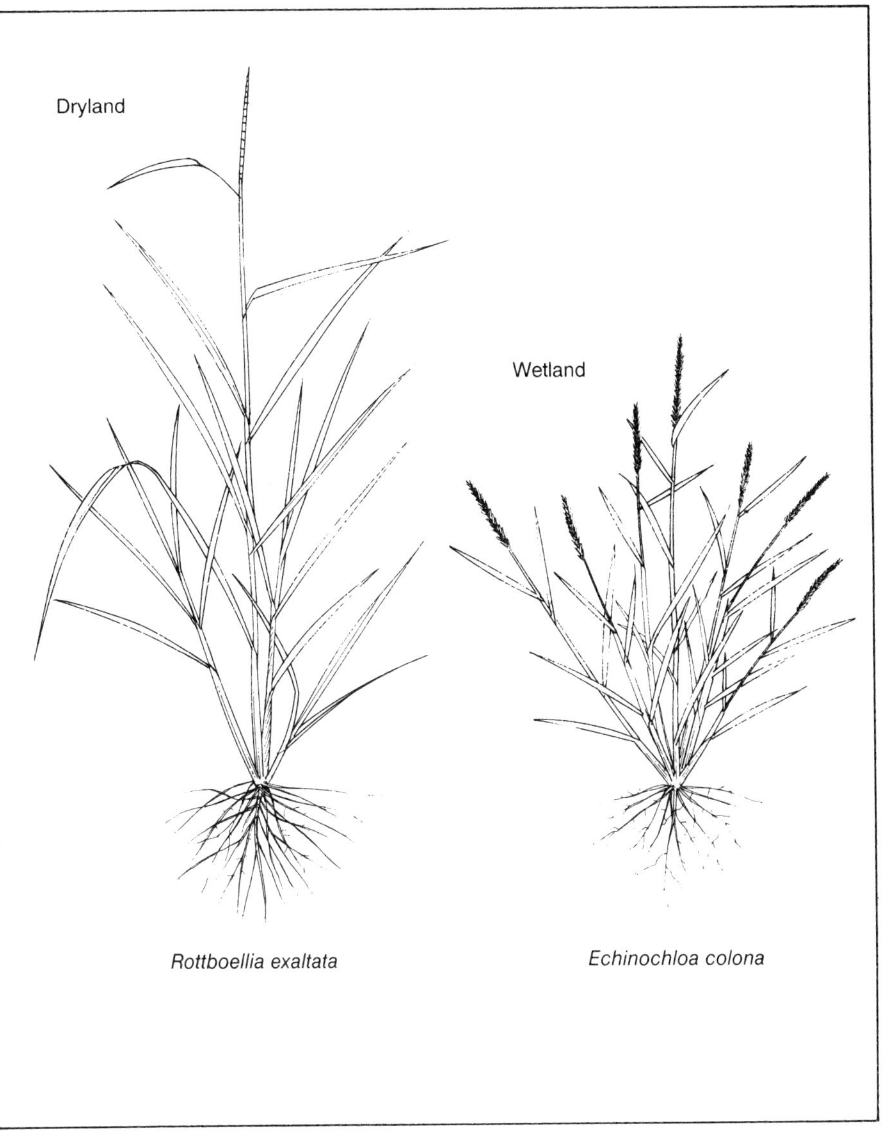

Rottboellia exaltata

Echinochloa colona

Common cowpea weeds — sedges

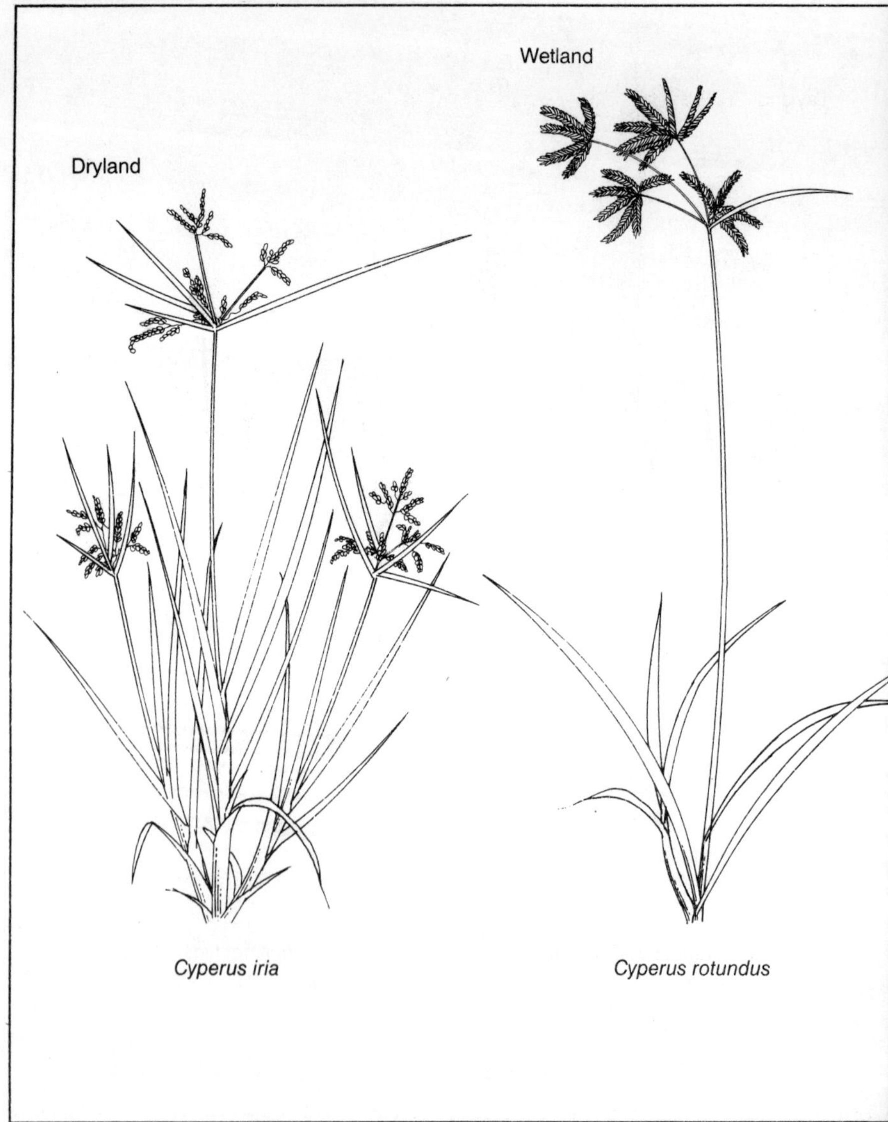

Common cowpea weeds
— broadleaf weeds

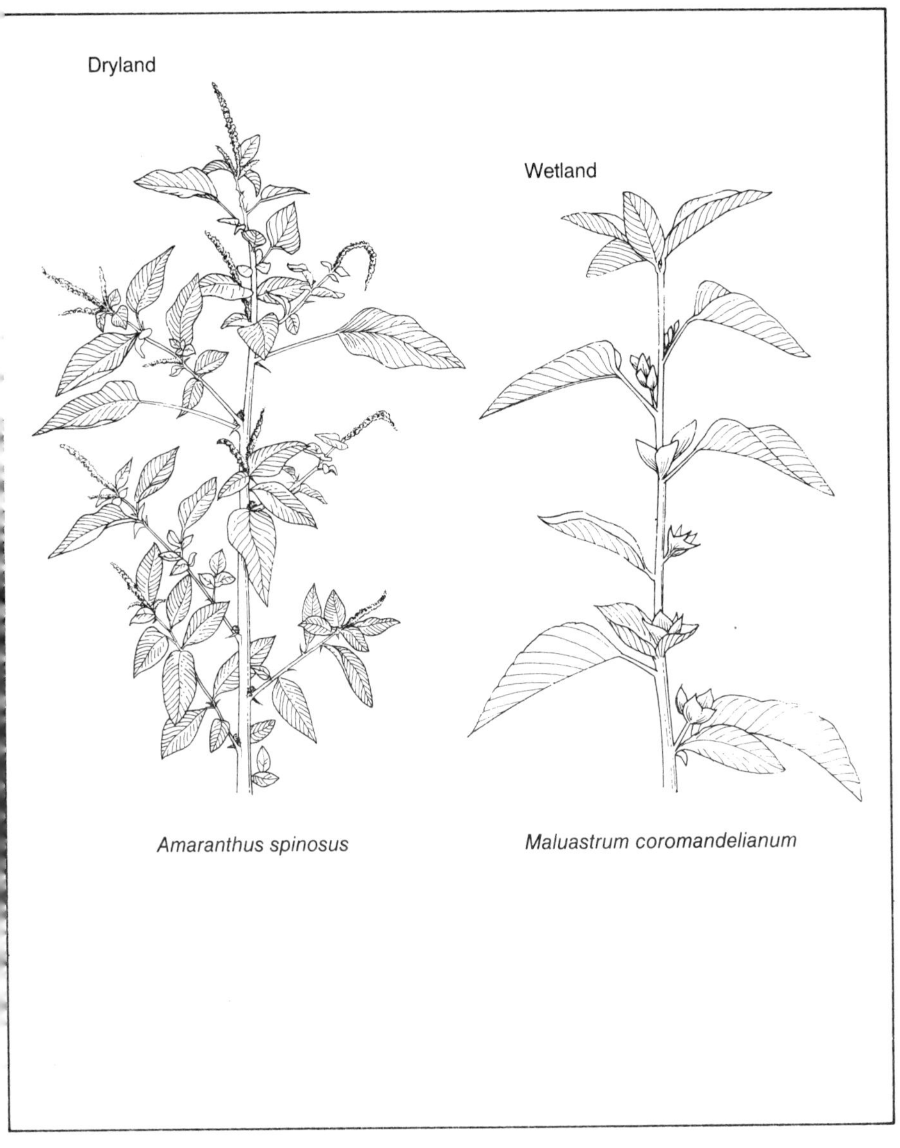

Dryland

Wetland

Amaranthus spinosus

Maluastrum coromandelianum

Yield reducers — insect pests

Yield loss to insect pests

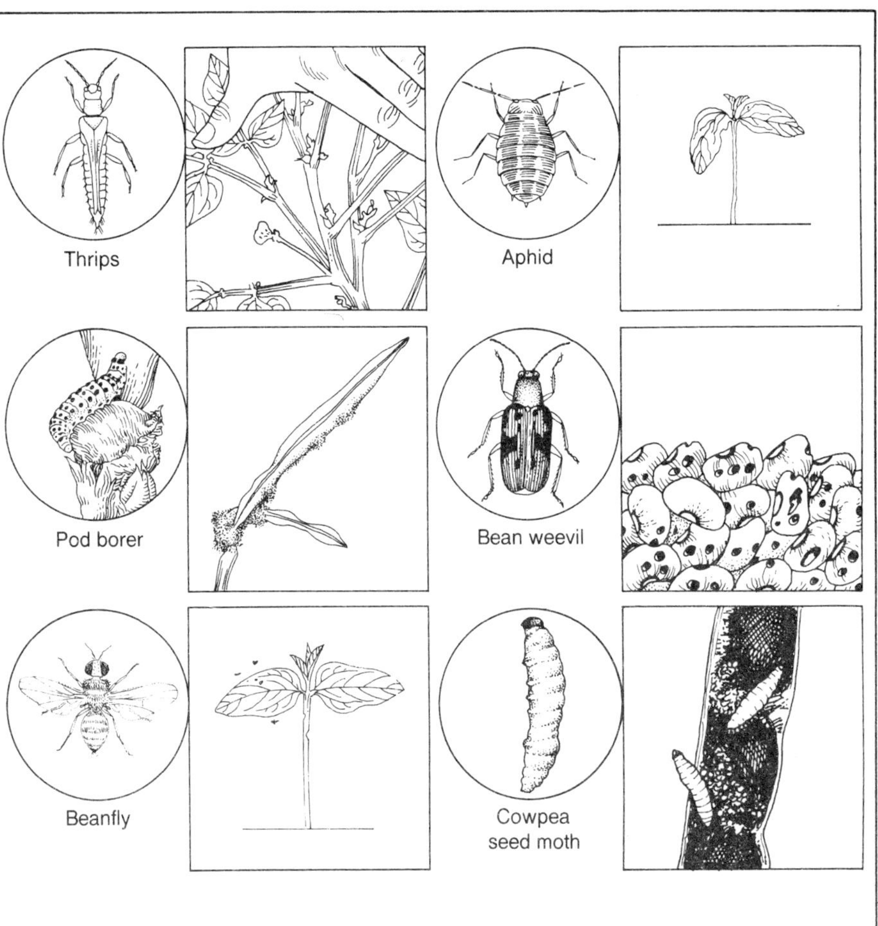

- Insect pests are a serious threat to the cowpea crop. They can attack all parts of the plant at all stages of growth.
- Uncontrolled insects can destroy the crop.

Controlling pests — using cultural practices

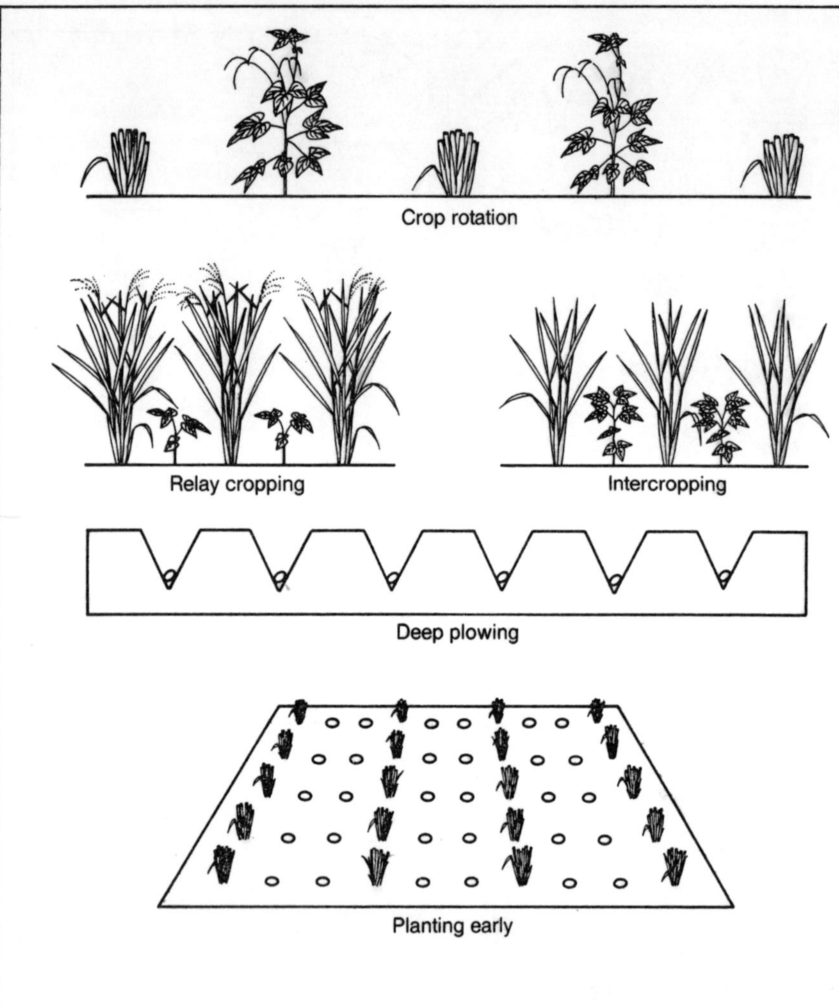

● Cultural practices can help reduce insect numbers.

Controlling pests — using insecticides

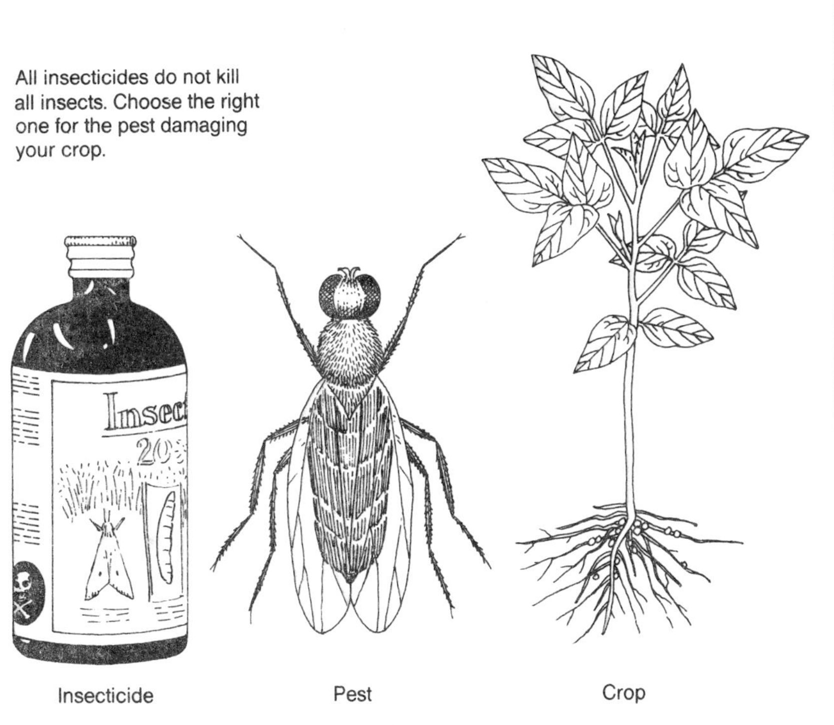

- Chemical insecticides effectively control many cowpea insects. Apply chemicals only as directed.
- Sprays are most needed at
  - 2 days after emergence
  - 12 days after emergence
  - flowering
  - 10 days later.

Controlling pests — planting resistant varieties

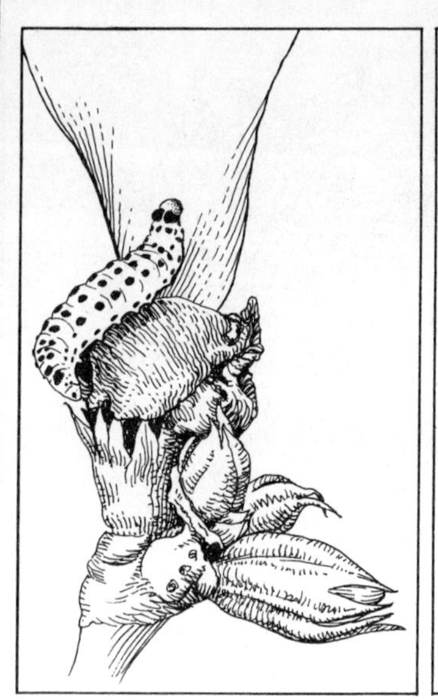

Not resistant

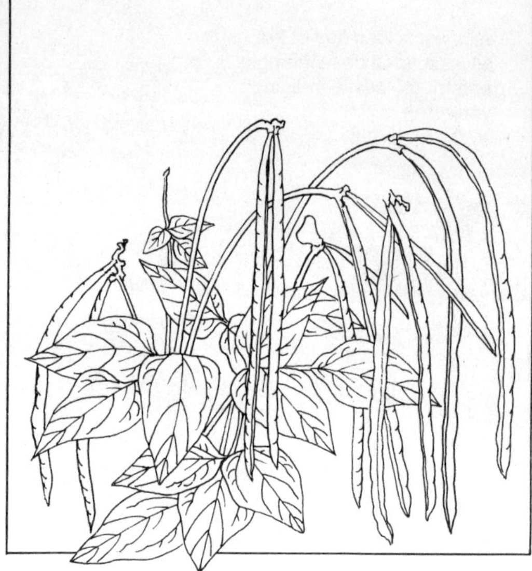

Resistant

- Some cowpea varieties resist pest damage better than others.
- Planting resistant varieties is a low-cost way of reducing insect damage.

Combining pest control methods

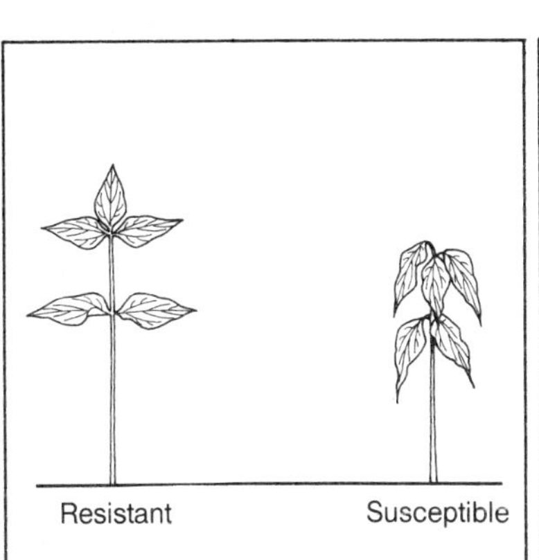

Plant resistant varieties

Use proper cultural practices

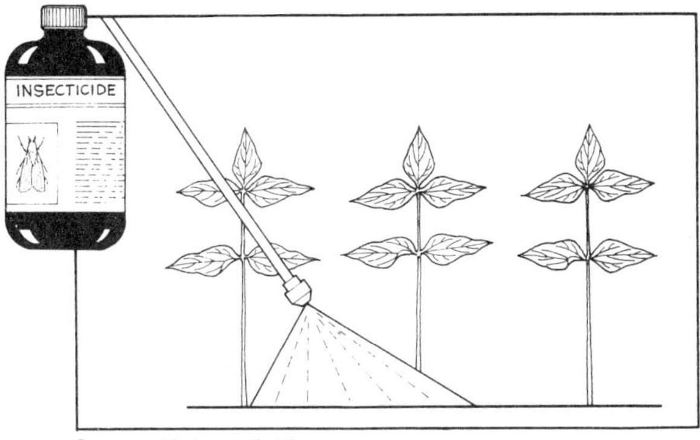

Spray with insecticides

- Several pest control methods can be combined:
  — using proper cultural practices
  — spraying the right insecticides at the right times
  — planting varieties that resist pest damage.

Common insect pests of cowpea — at seedling stage

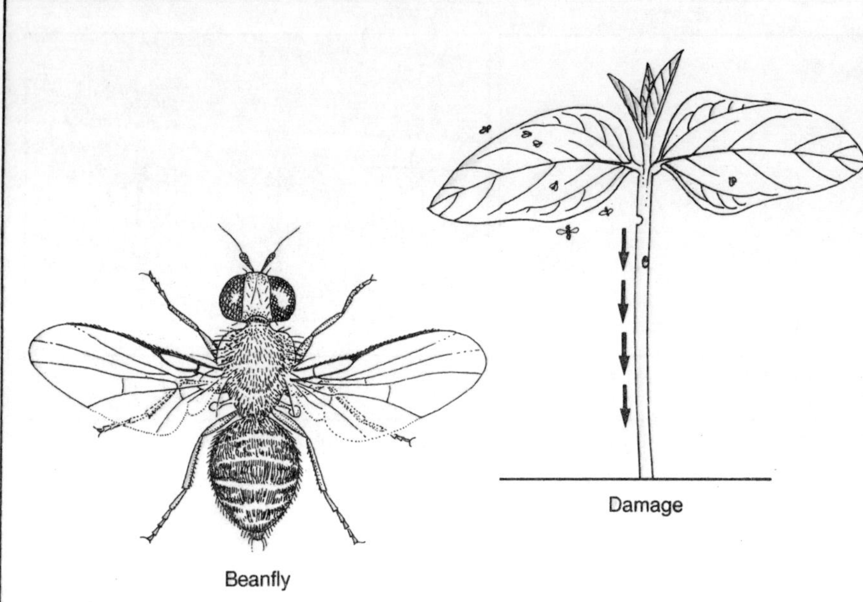

- Scientific name: *Ophiomyia phaseoli*
- Damage: The maggot bores into the stem and tunnels toward the base, damaging the stem. The plant withers and dies.
- Control: Plant varieties less susceptible to beanfly in your area. Spray seedlings with insecticide 2 to 3 days after emergence.

At preflowering stage

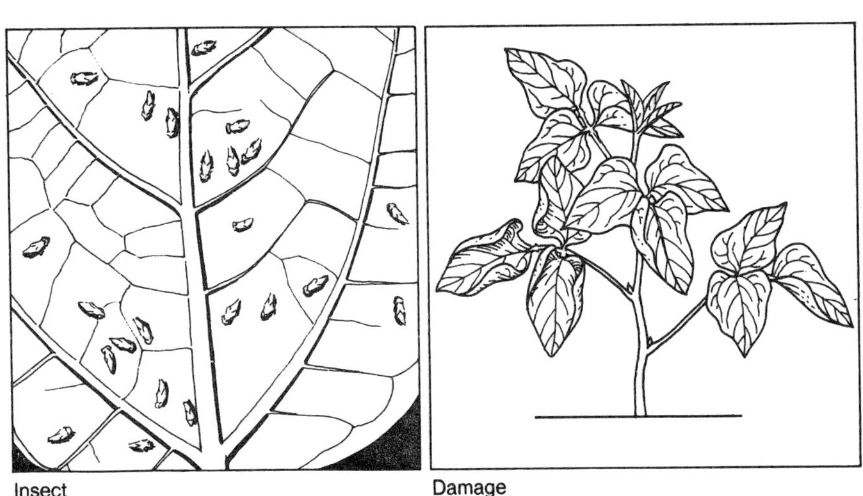

Insect

Damage

- Scientific name: *Empoasca* species
- Damage: Leaf turns yellow at veins and margins, then curls into a cup.
- Control: Plant varieties less susceptible to leafhopper damage in your area. Spray insecticide at preflowering stage, if needed.

At flowering

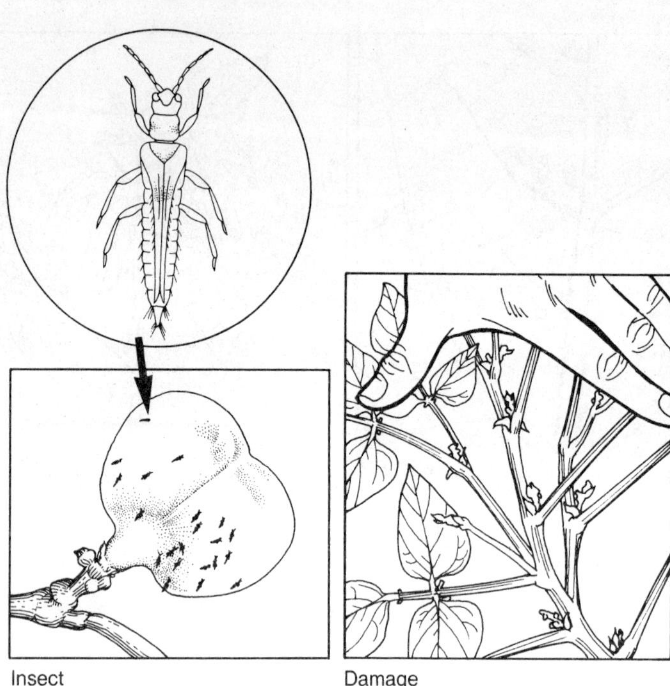

Insect

Damage

- Scientific name: *Megalurothrips* species
- Damage: Open flowers are distorted and discolored. They drop off and no pods are formed. When thrips are severe, plants do not flower.
- Control: Plant less susceptible varieties. Spray insecticide at flowering.

At pod formation

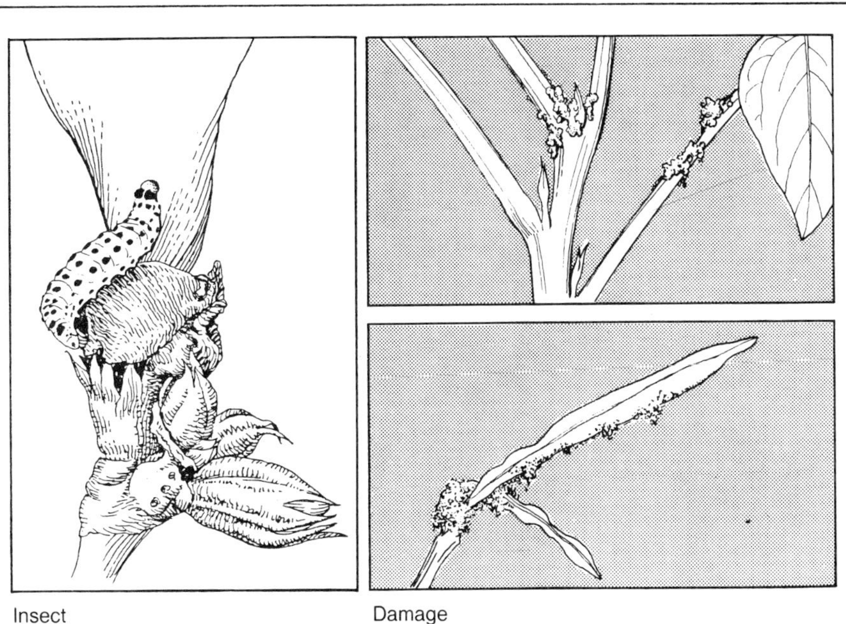

Insect

Damage

- Scientific name: *Maruca testulalis*
- Damage: Larva eats through leaves, flowers, and pods, leaving webbing and frass on them. Seeds do not fill.
- Control: Plant resistant varieties. Spray insecticide 10 days after flowering begins.

Preflowering to pod filling

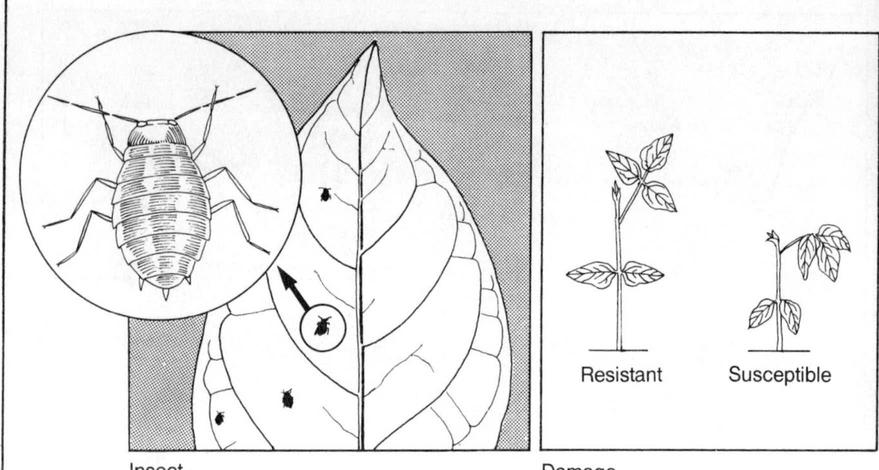

Cowpea aphid
Scientific name: *Aphis craccivora*

- Scientific name: *Aphis craccivora*
- Damage: Plant growth is stunted, leaves are distorted, and pods shrivel. No seed is produced. Aphids also carry cowpea mosaic virus disease.
- Control: Plant resistant varieties. Spray insecticide at pre-flowering stage.

Yield reducers — diseases

Yield loss to diseases

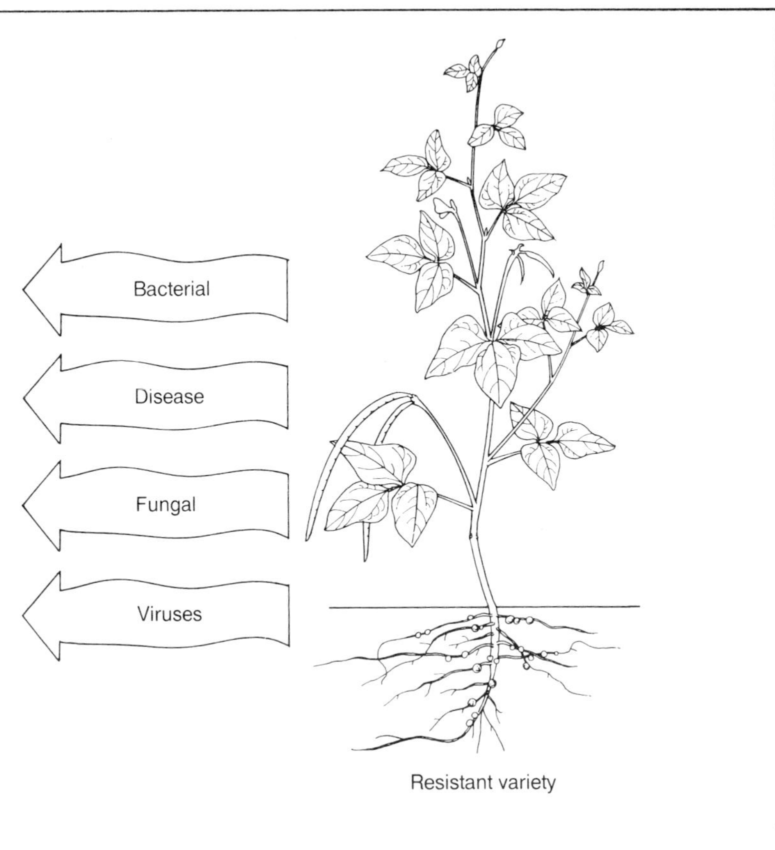

Resistant variety

- Fungi, viruses, and bacteria attack cowpea and can severely reduce plant stands and yields if not controlled.

Controlling diseases — planting resistant varieties

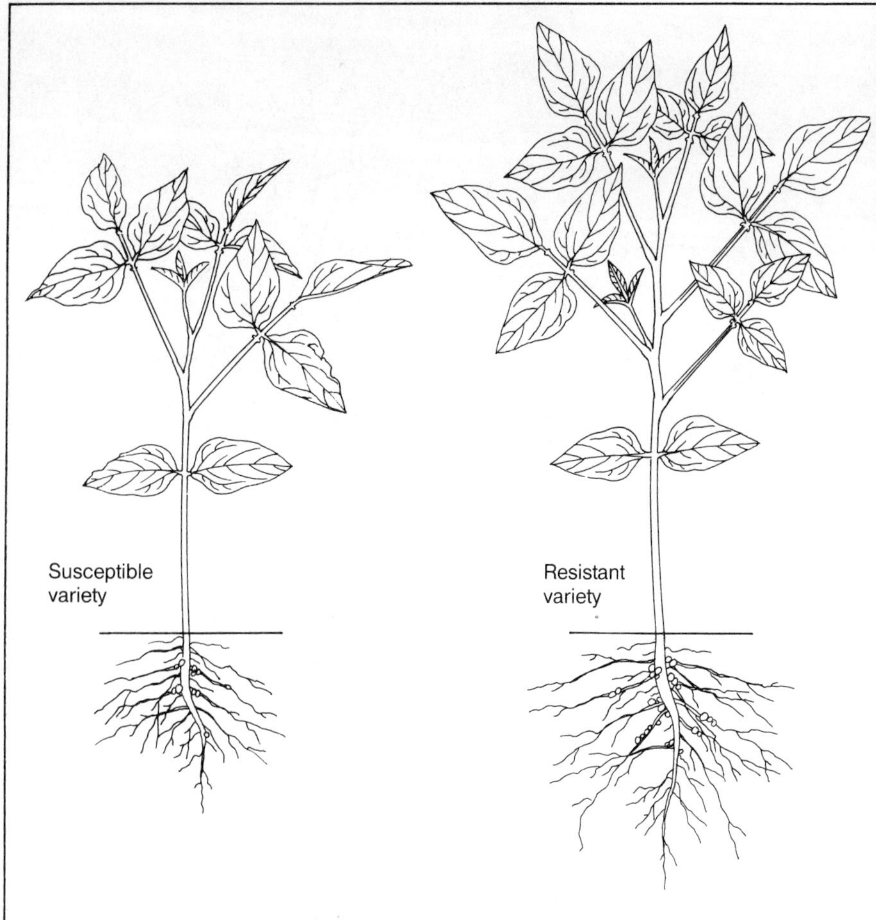

- Some cowpea varieties resist damage from certain diseases.
- Planting resistant varieties is a low-cost way of controlling disease.

Controlling diseases — using cultural practices

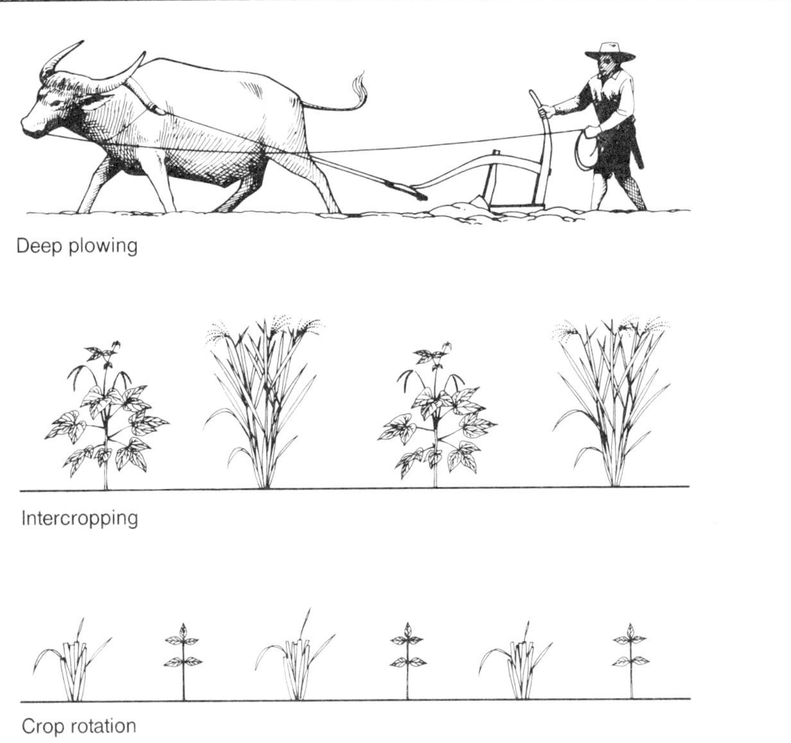

- Use cultural practices such as plowing, crop rotation, and intercropping to control diseases.
- Destroy crop residue because it may shelter and spread disease.

Controlling diseases — using chemicals

Treat seed with fungicide before planting

Seeds for planting

- Chemicals effectively control some diseases.
- To protect against soil-borne diseases, treat seed with fungicide before planting.

Common diseases of cowpea — Fusarium wilt

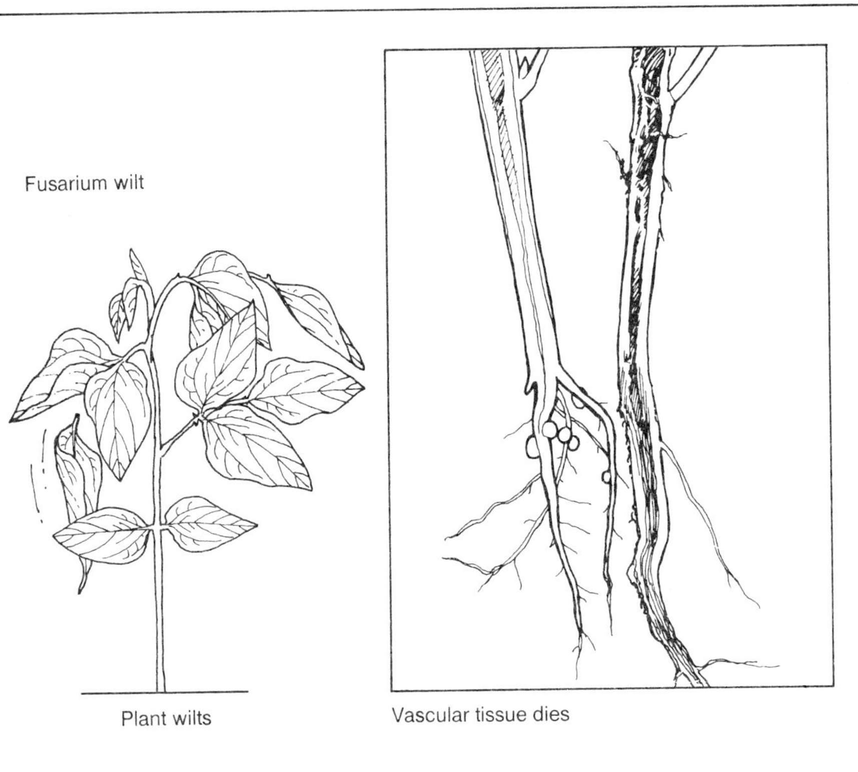

- Scientific name: *Fusarium oxysporum* f. sp *tracheiphilum*
- Symptoms: Leaves become limp and yellow, plants are stunted; young plants wilt rapidly, then die.
- Control: Plant resistant varieties.Treat seed with fungicide before planting.

Cercospora leafspot

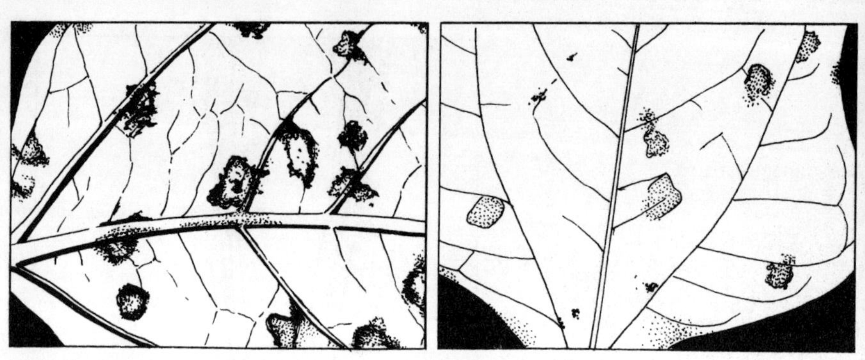

Lower leaf surface

Upper leaf surface

- Scientific name: *Cercospora canescens; Cercospora cruenta*
- Symptoms: Round or roundish cherry-red to reddish brown sores, up to 10 mm across, appear on leaves.
- Control: Use clean seed and plant resistant varieties. Treat with fungicide.

Brown rust

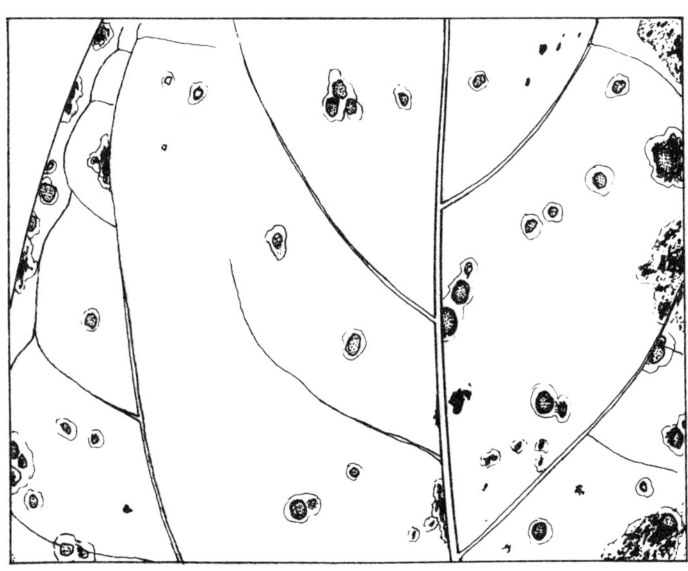

- Scientific name: *Uromyces appendiculatus*
- Symptoms: Blisters develop on leaves, releasing powdery, reddish brown spores.
- Control: Plant resistant varieties.

Brown blotch

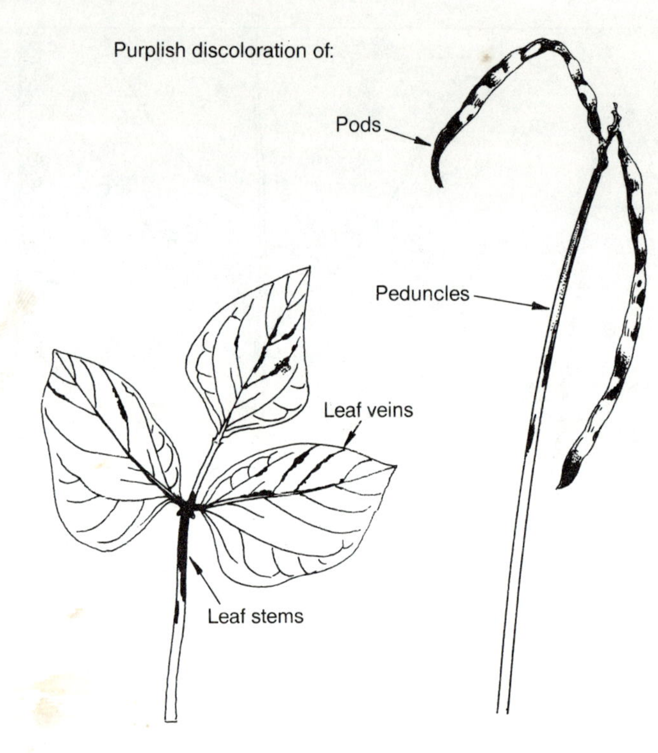

- Scientific name: *Colletotrichum capsici*
- Symptoms: Pods, leaf stems, and veins turn purplish brown. Flower stalks may crack. Pods twist and curl, do not develop.
- Control: Use clean seed. Plant resistant varieties. Destroy crop debris.

Powdery mildew

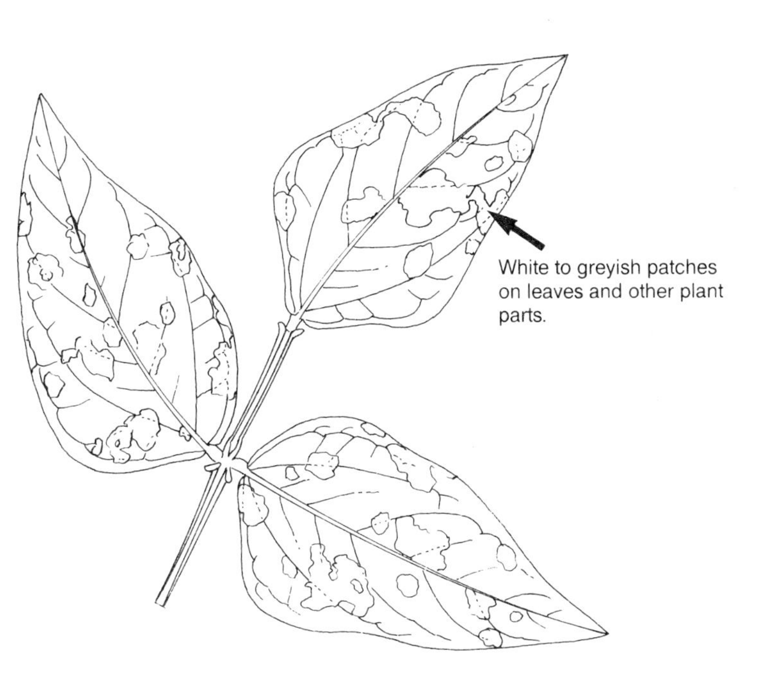

- Scientific name: *Erysiphe polygoni*
- Symptoms: White patches, turning greyish, and spreading on leaves and other plant parts.
- Control: Plant resistant varieties. Use fungicide.

Bacterial blight

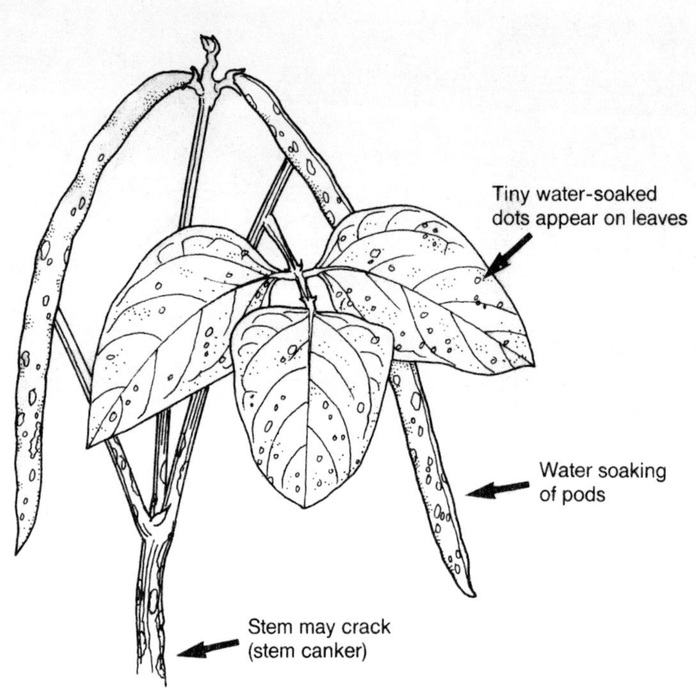

- Scientific name: *Xanthomonas vignicola*
- Symptoms: Tiny water-soaked dots appear on leaves; then the surrounding tissue dies, turning a tan to orange. Stems may crack and pods look water-soaked.
- Control: Use clean seed. Plant resistant varieties.

Cowpea (severe) mosaic virus

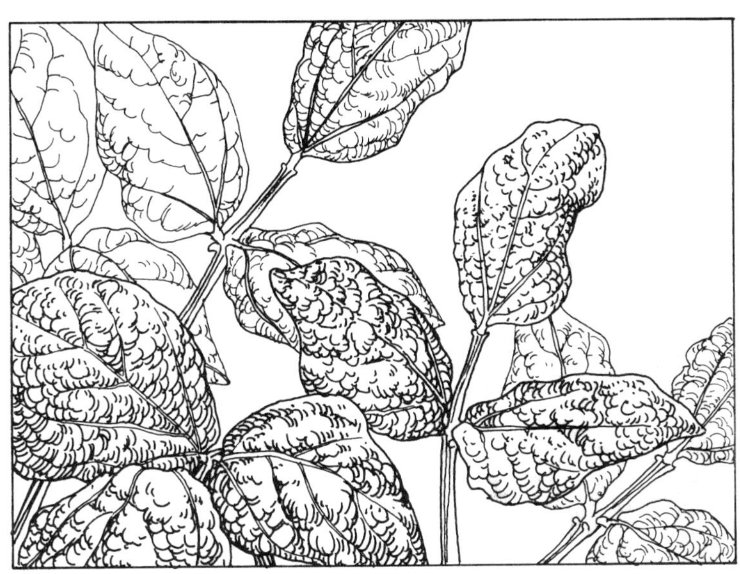

- Name: Cowpea (Severe) Mosaic Virus (CSMV)
- Symptoms: Leaves become mottled and distorted.
- Control: Use clean seed and plant resistant varieties. Control virus carriers such as beetles.

Cowpea golden mosaic

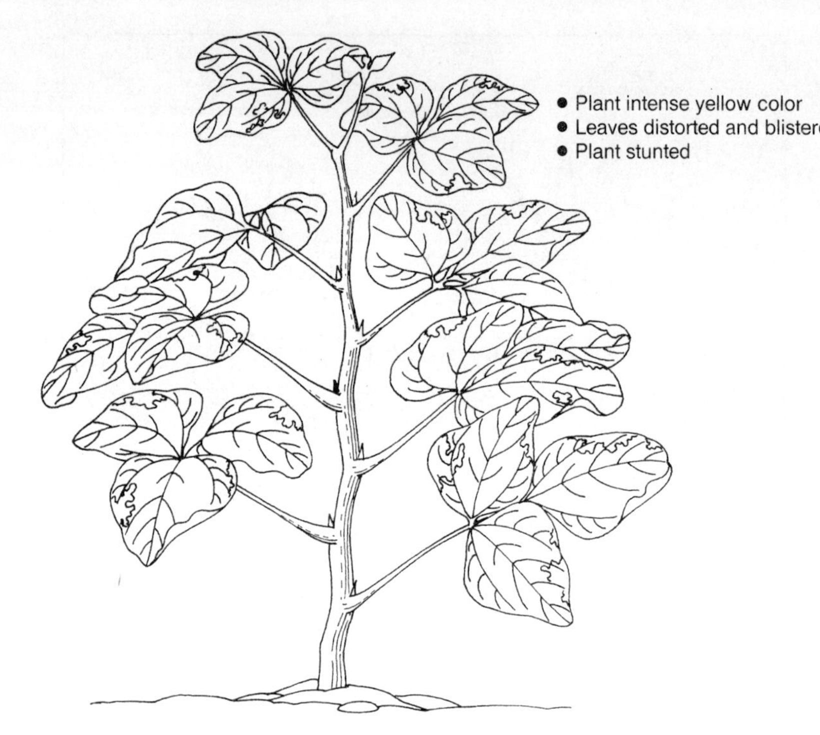

- Name: Cowpea Golden Mosaic
- Symptoms: Plants turn intense yellow; leaves become distorted and blistered; plants are stunted.
- Control: Plant resistant varieties; control the disease carrier, white fly (*Bemisia* sp.)

Cowpea in other cropping systems

Cowpea in other cropping systems — sequence cropping

Cowpea before maize

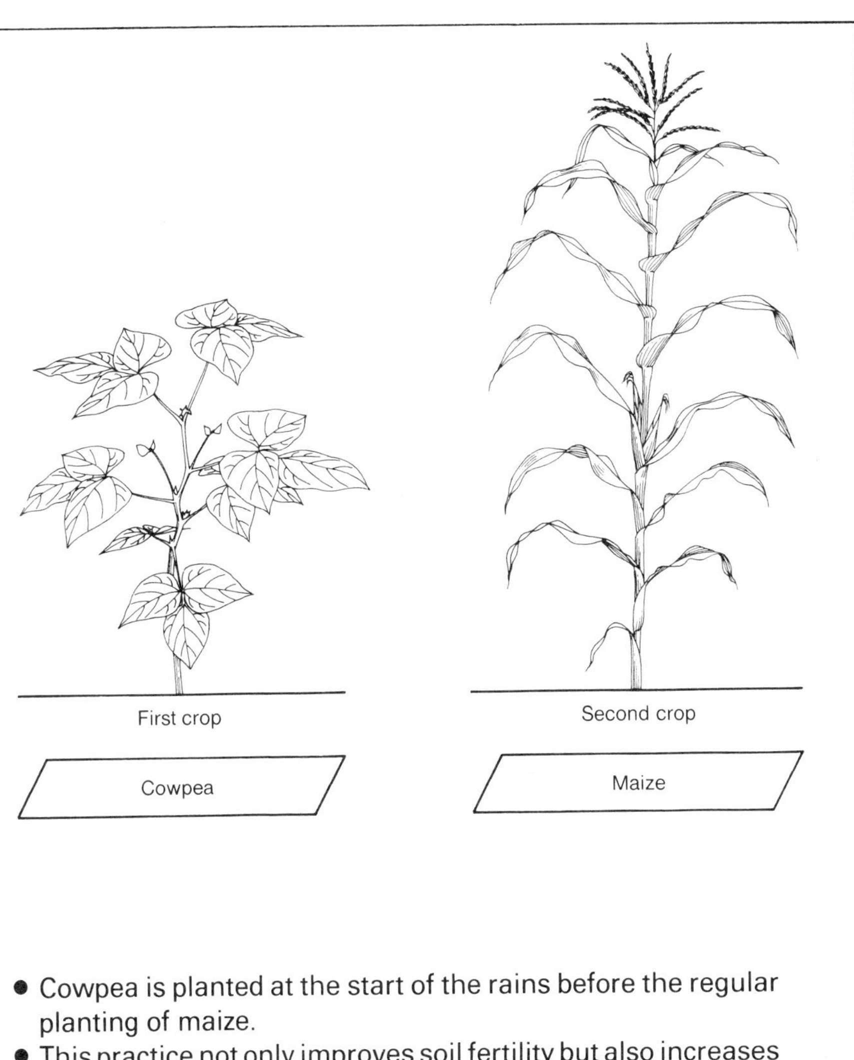

- Cowpea is planted at the start of the rains before the regular planting of maize.
- This practice not only improves soil fertility but also increases food production.

Cowpea before sorghum

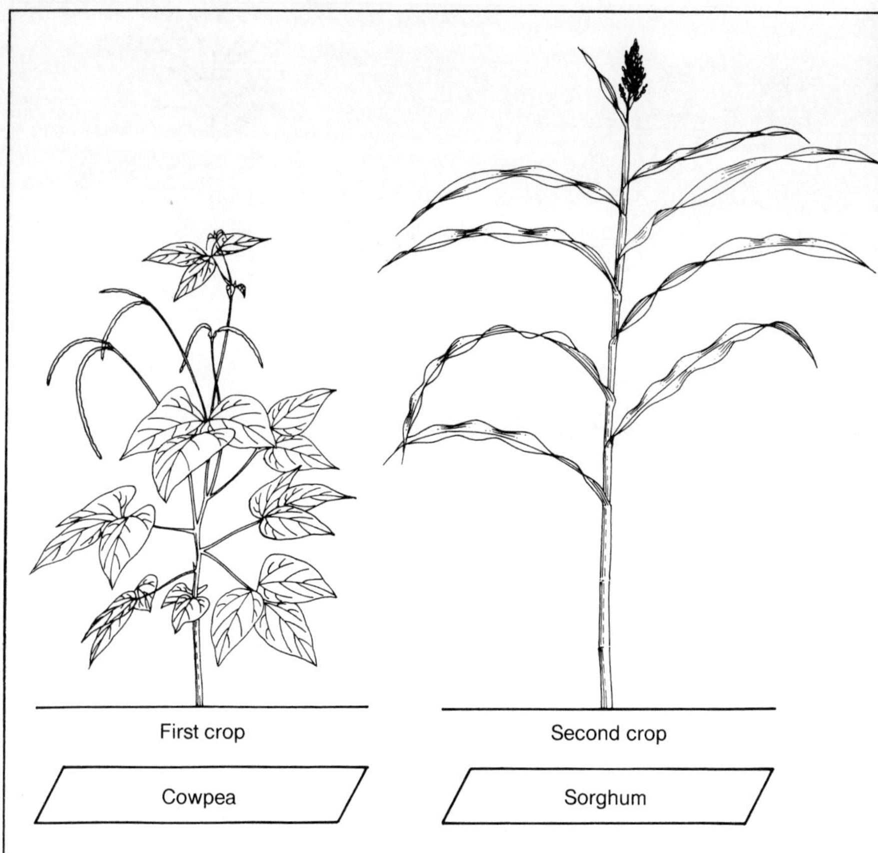

- Cowpea is planted at the start of the rains. Sorghum is planted after the cowpea harvest.

Cowpea before cotton

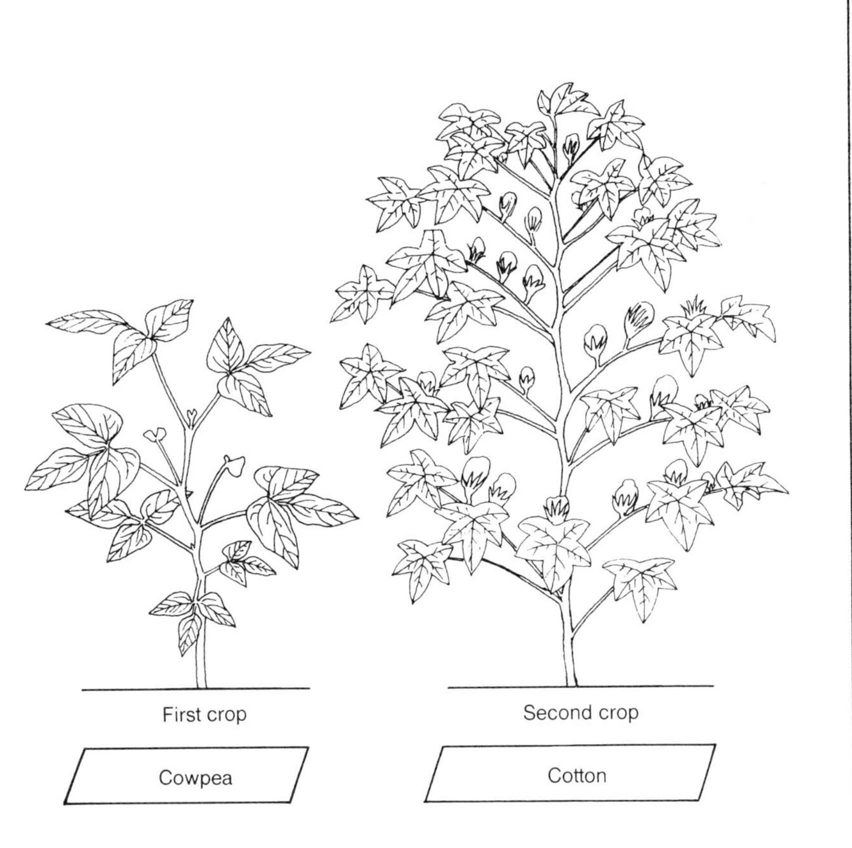

- Cowpea can be planted before the regular planting of cotton at the start of the rainy season.
- It provides additional income and food for the farmer.

Cowpea before wheat

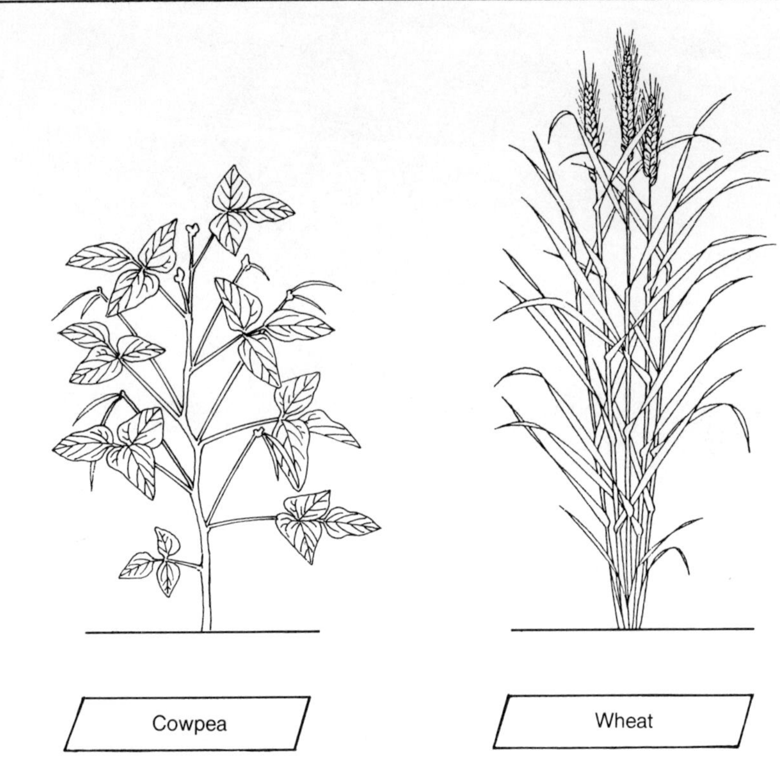

- The cowpea-wheat system can be practiced in subtropical Asia, where cowpea is planted in the rainy season and wheat is planted in winter.

Cowpea in other cropping systems — intercropping

Maize and cowpea

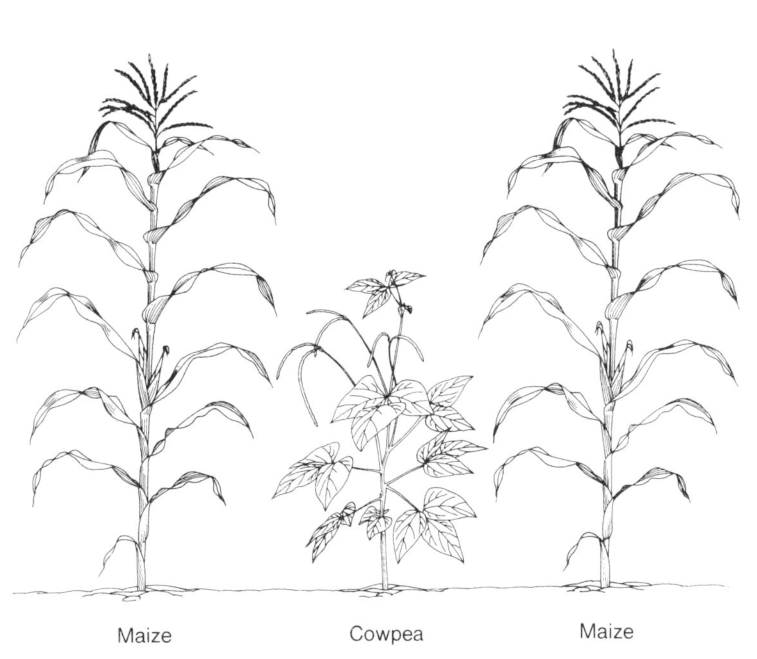

- Cowpea is planted between rows of the main crop, maize. Both crops are planted at the same time.
- This system insures against crop failure from drought and pests.

Sorghum and cowpea

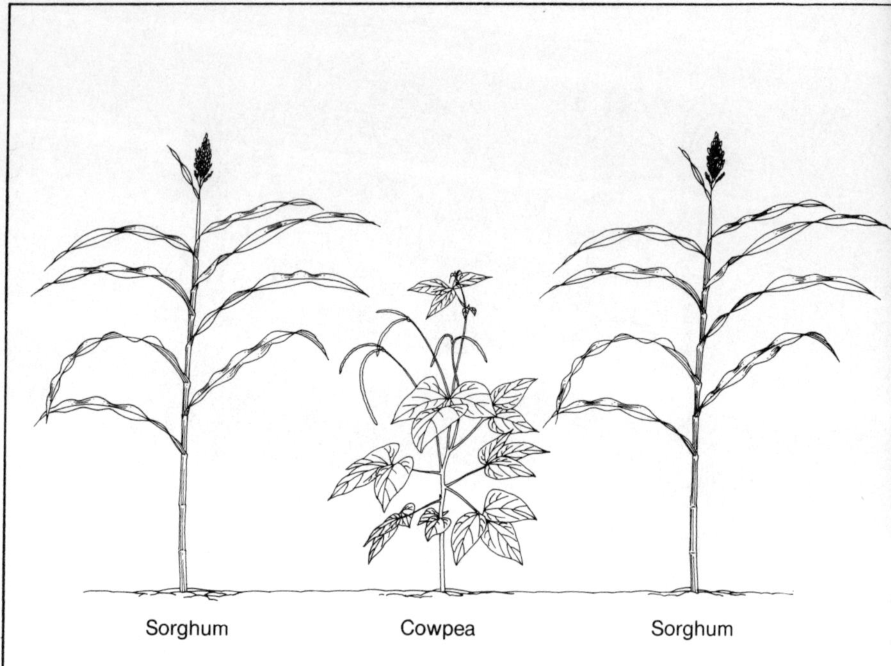

● Cowpea can be planted between rows of sorghum.

Sugarcane and cowpea

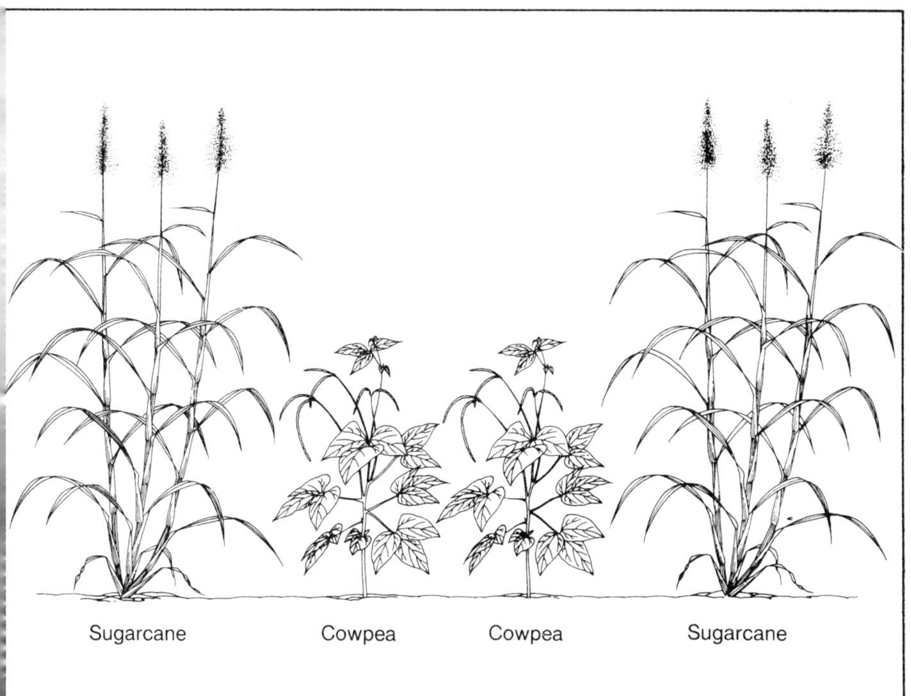

- When sugarcane is intercropped with cowpea, two rows of cowpea are planted between rows of sugarcane.

Cassava and cowpea

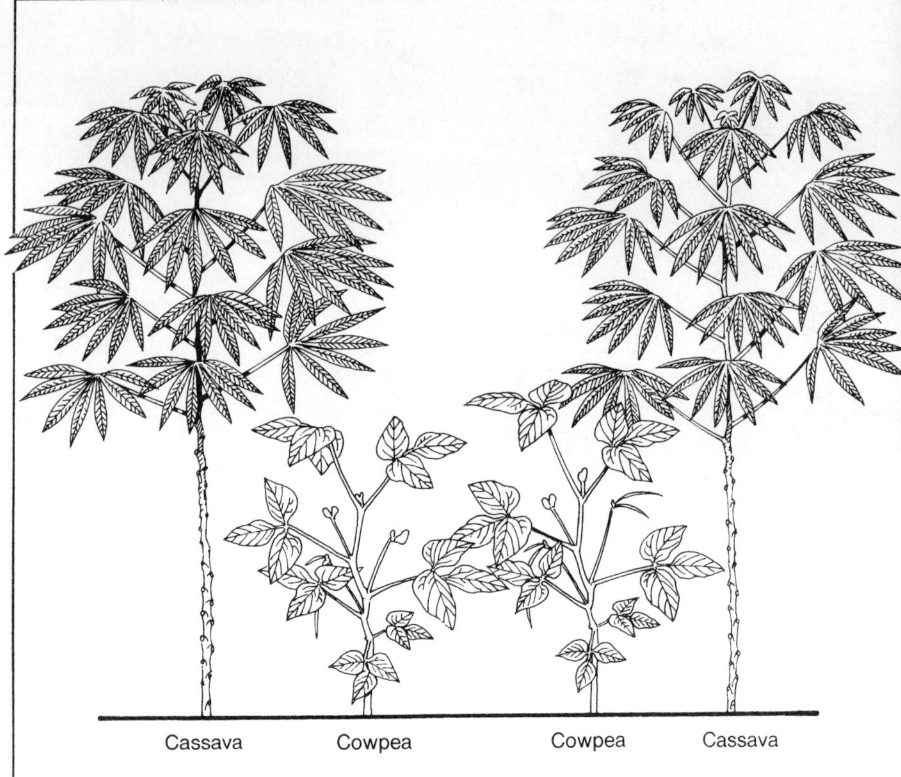

- Two rows of cowpea can be planted between rows of cassava.

Plantation crops and cowpea

- Cowpea can be planted in the vacant spaces of plantation crops.

Cowpea in other cropping systems — strip-cropping

215

Strip-cropping maize and cowpea

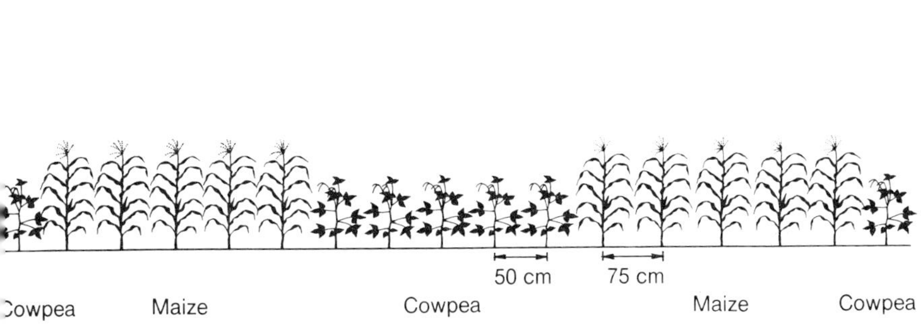

- Maize and cowpea are planted in strips of six to eight rows. Row spacing is 75 cm for maize and 50 cm for cowpea.

Strip-cropping sorghum and cowpea

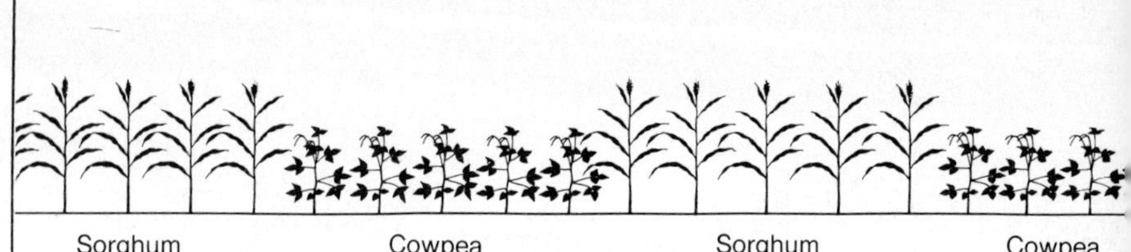

- Sorghum and cowpea are planted in strips of six to eight rows.